1979

The
Environmental
Crisis

The Environmental Crisis:

A Systems Approach

William James Metcalf

St. Martin's Press
NEW YORK

© University of Queensland Press,
St. Lucia, Queensland, 1977

All rights reserved. For information, write:
St. Martin's Press, Inc., 175 Fifth Avenue, New York, N.Y. 10010
Printed in Hong Kong
Library of Congress Catalog Card Number: 77–73798
ISBN: 0–312–25707–4
First published in the United States of America in 1977.

To my Mother and Father

Contents

Preface   xi

PART 1   ENVIRONMENTAL CRISIS   1

1. **The Idea of an Environment Crisis   3**

    Widespread acceptance of the idea of a crisis   3
    Critical questions   6

2. **Factors in the Environmental Crisis   8**

    Population   8
    Pollution   13
    Resources   14
    Agriculture and Food Supply   17
    The environmental crisis – Conclusion   18

3. **A Review of Proposed Solutions to the Environmental Crisis   19**

    1. Scientific solutions   19
    2. Tehnological solutions   21
    3. Economic solutions   23
    4. Philosophic solutions   24
    5. Social change solutions   27
    Conclusion   29

PART 2   A MODEL OF THE ENVIRONMENTAL CRISIS   31

4. **A Model for Understanding Environmental Problems and Solutions   33**

    The interrelated nature of environmental problems   33
    The interrelated nature of solutions to environmental
        problems   34

Systems model   35
Conclusion   40

5. **Relationship between Lifestyle and Ethical Subsystems**   42

Ethical determination of lifestyle   42
Lifestyle determination of ethics   44
Conclusion   46

6. **Relationship between Lifestyle and Technological Subsystems**   48

Technological determination of lifestyle   49
Lifestyle determination of technology   54
Conclusion   58

7. **Relationship between Lifestyle and Environmental Subsystems**   60

Environmental determination of lifestyle   60
Lifestyle determination of environment   64
Conclusion   66

8. **Relationship between Ethical and Environmental Subsystems**   69

Ethical determination of environment   70
Environmental determination of ethics   74
Conclusion   77

9. **Relationship between Ethical and Technological Subsystems**   78

Ethical determination of technology   80
Technological determination of ethics   82
Conclusion   86

10. **Relationship between Technological and Environmental Subsystems**   87

Technological determination of environment   88
Environmental determination of technology   92
Conclusion   94

11. **Self-induced Changes within Subsystems**   96

Self-induced changes in environment   96
Self-induced changes in lifestyle and ethics   100

Self-induced changes in technology 103
Conclusion 106

12. The Heuristic Model — Review and Observations 108

Stability and instability of the overall system 109
How to change the system — Environmental solutions 110
Conclusion 112

PART 3 APPLICATION OF THE MODEL 115

13. Application and Utilization of the Heuristic Model 117

Food supply and starvation as environmental problems 118
Reducing starvation and malnutrition 119
Application of the heuristic model 125
Conclusion 129

Notes to Text 129

Bibliography 135

Index 145

Preface

This book is based upon an M.A. thesis which I completed in 1975 at the University of Queensland. In rewriting the material for publication in book form, I have endeavoured to ensure that it can be understood by a wide range of people and that it is not a technical book; only chapter 4, which presents the theoretical basis for the system and model, might appear difficult.

For the sake of economy in this book, I have severely pruned the number of references and notes which were a necessary feature of the thesis. In deciding which to retain, I worked on two principles:— (1) information which appears questionable, and on which the argument rests, must be referenced;—(2) many less important bits of information came from interesting sources which I discovered during my two years of research and I hoped that a footnote for such less accessible material could save time for a reader using this as a source book.

PART 1

Environmental crisis

Part 1 contains three chapters which introduce the argument of this book. In chapter 1, I discuss the personal reasons which led me to undertake this work. The idea of an environmental crisis and how this idea is increasingly accepted, or at least acknowledged by many people, is also discussed. The problems of overpopulation, resource depletion, and pollution, as well as the composite problem of mass starvation, are briefly outlined in chapter 2. In chapter 3 many commonly suggested "solutions" to these environmental crisis problems are reviewed, and are categorized under five headings: scientific, technological, economic, philosophical, and sociological.

1 The idea of an environmental crisis

Like many other educated people, I arrived at the conclusion several years ago that there existed an environmental crisis. In order to become more knowledgeable about the dimensions of this crisis, as well as about how it might be corrected, I read from the usual environmental authors such as Paul Ehrlich, Barry Commoner, Rachael Carson, Gordon R. Taylor, Dennis Meadows, and René Dubos. But I found that while much of what I read confirmed my opinion that there was a crisis, many more questions were raised than answered. The authors argued and disagreed bitterly over many aspects of "the crisis", as well as over suggestions for its solution. In fact, it became obvious to me that not only were people arguing over details of the crisis but their modes of thought or their ideological perspectives differed so widely as to negate much of what purported to be rational argument.

It was this sort of confusion which prompted me to attempt to rationalize from the propaganda, to try to distil the facts from the opinion, and, most importantly, to try to understand in a logical fashion the nature of the environmental crisis. This personal questioning within the activist/conservationist/political circles I frequent led me to seek to produce a model or framework within which the masses of environmental propaganda to which we are all subjected could be critically examined. The presentation of this model constitutes the major part of this book.

WIDESPREAD ACCEPTANCE OF THE IDEA OF A CRISIS

It seems fairly clear that throughout Australia there has been a marked increase in environmental awareness, as well as acceptance of the idea that we face some sort of an environmental crisis. This trend has been indicated by several public opinion polls. For example, one poll, published in July 1974, reported that only 13 per cent of Australians think that "talk of dangerous overpopulation and pollution of our planet is the talk of cranks"; 57 per cent thought "unless controls are enforced quickly our world will become unlivable within 20 years"; another 30 per cent were undecided. Men and women as separate groups gave similar answers, though it was found that people's ages affected their outlook, with younger people being more concerned.

Another poll during 1974 basically agrees with these conclusions about the widespread degree of environmental awareness and concern, and found that this is not just a middle class issue, as is so often assumed, but is spread throughout society.

This trend toward environmental awareness has also been reflected by the rapid increase in environmental action groups throughout Australia, both in size and in numbers. For example, in Queensland alone, there are now more than seventy environmental action groups within the state-wide Queensland Conservation Council. This latter organization has changed from a volunteer group to a position of employing a staff of about ten in under two years.

Statements about environmental problems can even be heard from various industrialists. These statements of support for conservation, I have found, tend to follow a certain common pattern or format:—(a) we in X industry support conservation;—(b) criticism of X industry by environmentalists is too emotional, and is untrue anyhow;—(c) we in X industry are not nearly as guilty as Y or Z;—(d) anyhow, "society" needs us.

A good example of this form of apparently pro-environmental (defensive) argument is provided by Mr. G.P. Phillips, executive director, Australian Mining Industry Council.[1] (a) "The attitude towards the preservation of the natural environment which has grown up rapidly in the past few years is to be applauded." (b) "More than any other industry, mining has been the subject of intense criticism and propaganda from the powerful environmental lobby. Much of the propaganda has been emotional ... Cold, hard, and sometimes boring facts and statistics don't make good headlines." (c) " ... mining operations disturb only one hundredth of one per cent of the Australian land surface. For such an important activity, this degree of disturbance is negligible when compared with the disturbance to the environment caused by real estate development, farming, road construction, and other industrial development." (d) "But a rational approach to the problems of mining and the environment cannot be taken unless and until it is recognized that the principles of the preservation of the environment and the necessity to provide minerals for a growing nation are as important as each other."

In a similar vein:—(*a*) "Manufacturers are committed to pollution control and environmental protection ... they are concerned about any activity which contributes to a deterioration in the quality of life. The cleaning up of our environment is a desirable social goal, wholeheartedly supported by industry." (*b*) "Many Australians believe (incorrectly) that the cost of a clean environment will not affect them ... " (*c*) "An expanding population could be considered the fundamental pollutant of the environment." (*d*) "Environmental control, despite its high cost, has definite advantages for industry as well as for the community at large."[2]

These sorts of statements indicate a form of cashing in on, or jumping on, the environmental bandwagon. There are entire new industries, as well as subsidiary lines from established industries, designed to alleviate some particular environmental problem. Environmental protection may well prove to be the growth industry of the seventies, just as data processing was the growth industry of the sixties.

Even the politicians are jumping on this conservation bandwagon. In 1970 the then President Nixon declared himself a "conservationist". A special cover story in *Time* magazine was entitled "Environment: Nixon's New Issue". Jack Mundey, from the Communist party of Australia, has declared that the ecological crisis will remain the biggest political issue of the century. There are federal as well as state government cabinet portfolios for environment. Every politician, no matter how reactionary, states that he is concerned about the environment!

But while this noticeable upsurge in interest may be defined by some as awareness of ecological truths, others may rightly see in it elements of faddism. A quasi-religious movement, unfortunately, may have a scant logical basis. Also, such a movement is very much open to abuse and exploitation, because there is no logical basis for judgments. For example, "ecology jackets" were widely advertised in papers early in 1974. These, it was explained, were made from kangaroo skins and thus—somehow—they had something to do with ecology. "The sea need not fear man" is the punch line of an advertisement for plastic pipe used to pump raw, untreated effluent into the sea. There has even been a product called Eye-cology. This is a form of makeup which "enhances the natural environment around the user's eyes".

Such nonsense, unfortunately, may lead people to reject the whole environmental movement as equally nonsensical. But any movement which does not critically examine the premises on which it is built is subject to this sort of evangelistic fervour—and rejection.

It was my personal observation as an active, committed member of the conservation lobby that many environmental problems were being examined only within the confines of traditional academic subject areas. For example, air pollution was being studied by chemists, plant breeders, engineers, and even lawyers, each within an area of specialty, yet little interdisciplinary research was being done. Similarly, solutions to environmental problems were too often suggested in a simplistic fashion, reflecting the nature of the research. Also, interestingly enough, people often suggested environmental solutions in areas other than that of their own specialty. Thus the chemist might argue that air pollution could be controlled by the use of various economic measures such as taxes and subsidies, the lawyer might be confident that technology would provide a solution, and the industrialist might argue that more medical research would indicate pollution was no problem to health anyhow.

Thus, from far and wide I was assured that there was an environmental crisis, but the dimensions of this crisis were never very clear. There were many critical questions raised, and left unanswered.

CRITICAL QUESTIONS

It was this sort of unchallenged assumption which led me to enquire into the environmental crisis in greater detail. It was the observance of little critical interdisciplinary research which brought me to ask many questions, but the answers I found were rarely adequate, and often contradictory. Like anybody who becomes concerned about the environmental crisis, I desired to know what could provide a solution, if, indeed, there was a solution. My reading suggested to me many different solutions proposed by many different authors (chapter 3), who wrote either in ignorance of each other or in outright opposition.

I was left wondering how serious the population problem was. Was that *the* problem from which all other problems derived? Was modern technology the crucial problem? Could technology provide solutions? Was the environmental crisis a function of capitalism? Would a socialist revolution set things right? Was the Judeo-Christian heritage of Western man at the crux of the problem? Would Christian revivalism provide an answer? Are alternative lifestyles, group marriage, and other utopian experiments a solution? Or was any of the myriad suggested factors the crucial problem, or at the crux of a solution?

Not only could I not find adequate answers for these questions, but I could not even locate a common line of thought or underlying set of assumptions for all these ideas.

Gradually, I came to understand that the writings which troubled me the most were those which reasoned in a linear fashion. That is, a problem would be specified, causes located, and a solution suggested, all in a neat, unidirectional pattern. Almost invariably the reasoning would appear quite valid, since it did appear that these causes had led to (or certainly preceded) this problem, and that certain actions would solve or reduce the problem. But while each single linear argument appeared quite reasonable, there often arose a degree of contradiction. For example, the argument that misguided technology was the prime problem often contradicted sharply the argument that more technology would provide a solution. The argument that population growth was the problem, and must be controlled even by coercion, contradicted the argument that an authoritarian social system was the problem, and the solution lay in some form of countercultural freedom.

I then came to realize that the reason such linear thinking was so confusing to me was because it ignored the first lesson of ecology, i.e., that the world is a system. It is useless to look at any environmental problem and assume that simple causes and straight effects are particularly relevant. Also, it is naive to think that there is such a thing as a simple solution. Any intervention in a natural process will have many effects. To understand what these effects are it is necessary to utilize a dynamic systems framework (chapter 4). Natural systems, ecologists have told us, are examples of dynamic systems. To tamper with any natural system, therefore, will have complex results.

But to understand these results, or at least to know how to go about trying to understand the results, a framework for analysis was required. This I saw as a model within which the complexities of human ecology could be better understood. It was with the intent of discovering, explaining, and writing about such a heuristic model, that I undertook the research for this book.

2 Factors in the environmental crisis

In chapter 1, the general acceptance of terms like *conservation*, *ecology*, and *environment* was discussed. It seems that many people including academics, scientists, and lay-people agree that we face an environmental crisis.

Yet what are the environmental problems? What is the crisis? Is there one crisis, or are the problems too diffuse to be lumped together? How much of what is called the environmental crisis is based on scientific observation, and how much is based on aesthetic judgments? Why a crisis now, when man has existed on earth for thousands of years? What are the time factors which lend immediacy to the crisis?

These are some of the questions which I shall discuss in this chapter. Yet, such issues are so broad that at best I can only provide an overview at this stage. For example, there is material written every day on population, resources, and pollution. Let us therefore just consider the types of problems which are the subject of this book.

The discussion will cover the three primary problems of population, pollution, and resources; the acute problems of agriculture and food supply; the interrelationships between these problems; and finally the time dimension which shifts such problems to a crisis position.

POPULATION

In many ways, population seems to be the simplest, most easily understood of the several environmental problems. It is much easier for people to grasp the difference between 4 billion and 16 billion people than to understand the threat to the earth's ozone layer from supersonic transport (S.S.T.) flights, or the danger to the oceanic ecosystems from the use of DDT, for example. Population, because it is so personal—we and our children—has long been a very emotional issue. The media have given fairly wide coverage to over-population prophets such as Paul Ehrlich, as well as to population concern groups such as Z.P.G. In Australia, the National Population Inquiry of 1973–74 is indicative of this interest.

It has been my observation that warnings of overpopulation and population explosion were first dismissed by many people as alarmist, but more recently have come to be accepted by many people as *the* problem. To dismiss population growth, or to see it as *the* problem, are equally counterproductive, since neither help to solve environmental problems.

The current population concerns can safely be dated as starting from Thomas Malthus who argued in 1798 that as population increased geometrically (1–2–4–8–16–) and food supplies increased arithmetically (1–2–3–4–5–), mankind was bound always to teeter on the brink of starvation.

This "dismal theorem" seems to have been avoided for quite some time. Expansion of the frontier in the European colonies provided a much wider resource base for food production. Thus, for the Europeans at least, Malthusian predictions of starvation were not fully realized. Over time the birth rate declined for most sections of the population (first European, then more recently in other countries). This declining birth rate led to some fears during the 1920s and 1930s of a gradual global depopulation by humans. If the trends of decreasing birth rates continued, some people argued, mankind would fall below a replacement level of births, and his numbers would start a long, slow decline to extinction. Such fears are still occasionally expressed, though without any factual foundation.

These fears of gradual depopulation, generally based on an observed declining birth rate, have proved wrong. Longevity, fewer infant deaths, and fewer deaths from most communicable diseases all combine to increase the world population, at an ever increasing rate, in spite of the decline in birth rates. Over the past two hundred years, the world's population has been increasing at an exponential rate. Figure 1 illustrates the nature of this increase.

Currently, the population of the world increases at approximately 2.1 per cent per year, doubling every thirty-three years. There is an extra 76 million people annually, an extra 200,000 people every day, or 150 extra people every minute. The population of Australia is added to the world's total every two months. While it has taken millions of years for mankind to reach a population of 4 billion, the next 4 billion will be added before the year 2010, well within the lifespan of most people alive today. The United Nations Environment

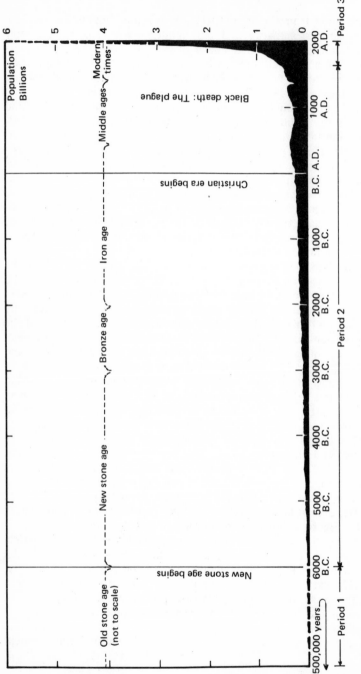

Fig. 1

Protection Group, in May 1975, predicted that the world's population may have a doubling time of only twenty-five years, if current trends continue.

There have been numerous predictions of the future population of the world. A study at Massachusetts Institute of Technology indicates that world population could reach 12 billion by 2033, 24.8 billion by the year 2066, and 48 billion in 2100. But these figures are based on a number of assumptions about human behaviour—always a difficult thing to accurately predict.[1] The United Nations, for example, has five possible projected population growth predictions, depending on which set of assumptions holds true. Three of the United Nations predictions are shown in figure 2.

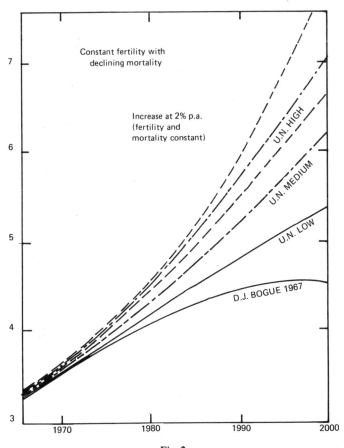

Fig. 2

Many people attack such predictions, pointing out how wrong they have been in the past. For example, world population today stands well above the highest curve on the population prediction chart published by the United Nations in 1964. One author[2] has argued that all population forecasts, without exception, have proved to be too low, even when made by people anxious to prove the existence of a population explosion. For example, Fairfield Osborn of the United States Conservation Foundation, in *Our Crowded Planet*, published in 1963, warned that the population of the United States of America could reach a figure of at least 190 million by the year 1975. He doubted it would be 200 million by the end of the century. Yet in reality, the United States population reached the 200 million mark in 1969, only six years after his dire predictions!

But while there is still considerable debate over the probable growth rate of population as well as the effectiveness or futility of various methods of population control, there is a generally perceived belief that there is a population problem. There are few writers who will argue that population increases, on a global basis, are anything but bad. But why is population growth seen as a problem, and why is rapid population growth seen as an even worse problem?

That there is a direct relationship between increases in population numbers and environmental impact is undeniable. More people require more food, utilizing more marginal land, more agricultural chemicals (given current agricultural economics) and therefore resulting in more soil destruction. More minerals, water, and energy are also required to maintain living standards. Below some critical level, natural systems such as a watercourse can cope with and break down pollution. American pioneers had a saying that a stream purified itself every ten miles, and there would be some truth in this as long as there were few enough people, and little enough pollution. But beyond a certain critical point, the natural purifier breaks down. This may also apply to the air cleaning properties of forest areas.

While a direct relationship between population and environmental destruction is generally accepted, there is disagreement over the numerical basis of the relationship. A famous debate between Paul Ehrlich and Barry Commoner, two environmentalists, has centred around this question. The disagreement does not question that population increase is environmentally harmful, but it does concern to what extent the environmental crisis is a result of population increases. This debate recurs at several points in this book, so will only be touched upon here.

Commoner argues that $I = P \cdot F$ where I is the environmental impact, P is population, and F is a function which measures the per capita impact. According to this equation, if population increases by

2 per cent, the environmental impact will likewise increase by 2 per cent. But while the United States population is currently growing at about 1 per cent, the environmental impact is increasing at a much faster rate than this (about 5 per cent). Commoner therefore concludes that while population growth is a problem, it is a relatively minor one.

Ehrlich, on the other hand, asserts that the $I = P \cdot F$ equation only holds for simple technology, low-population situations; that, in fact, the relationship between environmental impact and population in the contemporary context is $I = P \cdot F(P)$, signifying that the environmental impact per capita increases as the population increases. This is, in a sense, a case of diminishing returns, or overloading of natural systems. Ehrlich thus reasserts that population growth is a major, if not *the* major environmental problem, though he agrees it is one of several problems.

POLLUTION

Pollution would perhaps be the one environmental problem which everyone, to some extent, accepts as a problem. But the general acceptance is often at the expense of limiting the definition of what pollution is. It is not uncommon, for example, for pollution to be confused with litter, and both to be confused with conservation.

While it would seem safe to assert that most people would see litter as bad, other forms of visible pollution are not always appreciated as such, and invisible forms of pollution are even less well understood. It is not uncommon for people in developing areas to see smoke from a factory not as pollution but as a sign, or "smell" of progress. The haze which settles over a city for much of the time may become regarded as an almost natural feature of the city, rather than a concentration of dangerous chemicals.

Because there is so much being written about pollution, and because there are so many sources and types of pollution, as well as environmental and health effects from pollution, this section of this chapter can, at best, provide a very brief picture. It is doubtful if even the most hardened critic could deny that pollution exists and is a problem, though the crisis nature of this problem is hotly debated.

There are many cataclysmic environmental effects which, scientists argue, may result from pollution. For example, agricultural, industrial, and organic pollutants have been dumped into the Great Lakes system over many years. In Lake Erie, this has raised the Biological Oxygen Demand to a point where most of the life within the lake has ceased. A similar process is under way in the oceans of the world, where most sewage eventually settles. This will greatly reduce fisheries production, and may already have done so. But also, since much of the world's oxygen supply is provided by aquatic plants, any threat to the well-being of the oceanic ecosystem directly threatens the oxygen supply on which we are all dependent.

There are various cataclysmic effects predicted to result from serious air pollution. A heating up of the world (greenhouse effect) as well as a cooling down of the world (ice age effect) have both been predicted by various scientists. These are two quite possible, yet divergent tendencies, and the environmental effects of either could be disastrous. Sea levels would change, deserts form, cities flood or be left on dry plateaus, etc. While no one is quite sure which way things will go, particularly given a naturally shifting climate in any case (chapter 11), there is quite general scientific acceptance of global climatic dangers resulting from severe air pollution.[3]

Another cataclysmic effect of air pollution results from the exhaust gases from supersonic transport planes, such as the Concorde. The exhaust is in the ozone layer of the atmosphere, and reacts with the ozone to decrease its ability to filter out harmful ultraviolet rays from the sun. Though there is still considerable debate among scientists, there are many warnings that increased use of planes such as Concorde will permanently increase the harmful solar radiation on earth, leading to marked increases in skin cancer.

Thus we find that pollution, no matter at what level we wish to look at it, is environmentally harmful. There are many ways in which ecosystems are disrupted by even relatively minute quantities of pollutants such as mercury or DDT. Also, there are many direct threats to human health from such pollutants. Research into this goes on all the time, and new health threats from pollution are regularly announced. Using statistical techniques somewhat analogous to those showing a link between cigarette smoking and cancer, researchers are finding new correlations between various pollutants, and both health factors and death rates. "Pollution kills" is no longer just a conservationist's catch-cry.

RESOURCES

At any one time a culture has a range of resources which can be used for productive purposes. Petroleum, coal, various minerals, as well as manpower, technical skills, and a stable political climate might all be defined as resources. One basic criterion for separating such resources however is their renewability. Manpower and technical skills are renewable resources and technology in particular, tends to increase through usage, rather than decrease (see chapter 11). Some energy sources such as solar, tidal, and hydroelectric are renewable, whereas other energy sources, such as coal and gas, are basically nonrenewable. Minerals, of course, are nonrenewable, in that only a finite amount exists.

Nonrenewability of resources is different for compounds such as coal, and for minerals such as copper and iron. When coal or gas is burned, the particular configuration of carbon atoms is broken down, releasing energy and smoke. The carbon is not lost, but is in a lower energy state. The energy gained from the coal eventually becomes heat and is dissipated, and the resource is truly destroyed. Coal and petroleum, of course, are renewable over very long periods of time, but because of the rapid rate at which we now use these, for all practical purposes they may be considered as finite and non-renewable.

For minerals such as copper, iron, or gold, a finite amount exists now, has always existed, and for all practical purposes, always will exist. These resources are not really used up in a strict sense. But what does happen is that the finite amount of resource is spread over the earth in various concentrations, with the more diffuse amounts requiring greater energy to extract. As mankind uses more of the finite amount of a mineral, ever more energy must be expended to extract the next unit. It is basically the energy limitations, for whatever technology is then used, which limit the availability of a mineral resource.

But once the mineral is extracted, it is used in some form of process and then dispersed over a wide area. Chemical reactions such as rust will often occur, changing the mineral to a compound. This whole process is referred to as entropy, the tendency to run down. Thus, there occurs entropy when iron is turned into steel for cars, which gradually turn to rust and decay. At any point, the iron could be reformed, but relatively greater energy for its extraction would be needed.

Because of the renewability of certain resources, the ever-existing state of mineral resources, though widely dispersed, the changing technology of resource mining and recycling, and in fact the changing recognition of what are and are not resources, it is difficult at any one time to estimate how much of a resource exists. Many assumptions must first be made. Also, there is the fact that many resources are still being discovered in new places in the world, though here again the new sources are generally more difficult to harvest, (e.g., North Sea oil), the easiest sources having been tapped first.

According to estimates collected in 1972, if we continue to use the known reserves of coal at current rates, it could last 2,300 years, but if our coal usage continues to increase at the current exponential rate, we only have enough for 111 years. Even if known reserves are wrong (which is quite likely) and there exists five times as much coal as is currently known, this would add only an extra 39 years' supply. Similarly, aluminium could last 100 years at current usage rates, but under exponential increases of usage would last only 31 years, and five times as much aluminium would last only an extra 24 years. Thus, we find that the exponential rate of increase of usage of these resources is one of the prime factors changing a situation of limited resources to one of resource crisis.

There are many people who attack these sorts of figures and the ensuing dire predictions. The Australian economist Colin Clark, for example, argues that had such "known reserves" in 1949 been correct, then the world ran out of lead and zinc some years ago, and should have run out of copper in 1972. Just as the 1949 known reserves were wrong, so too are today's known reserves wrong. Therefore, Clark argues, we shall always have enough resources, the very amount of known resources at any one time being determined by the amount of exploration, and the perceptions of scarcity.

John Maddox,[4] the past editor of the prestigious journal, *Nature*, has also attacked this sort of scarce-resource reasoning. He argues that a great deal of the world has not yet been explored for minerals, and (perhaps) there is probably much more to discover. Also, technological developments will improve mining methods so that minerals which in the past were not worth extracting may become usable. Also, as a mineral becomes more scarce, its price will rise, and this increases the pressure on scientists and industrialists to devise resource substitutes, as well as encouraging the recycling of resources.

While no one would deny that most likely there are undiscovered sources of most resources, and many would grant that these sources may be relatively large, the undisputed fact still remains that all resources are finite. While technology, most likely, will be devised to enable mining of more dispersed minerals, and will enable some resource substitution, this is only a stop-gap measure. Given our current rate of resource usage, and the exponential rate at which our resource usage increases, no matter how much exists we will one day run out. And given reasonable predictions of perhaps to be discovered future resource supplies, the date at which we shall run out is not so far removed.

AGRICULTURE AND FOOD SUPPLY

This is not only an environmental problem, but also a result of certain environmental problems, and the cause of other environmental problems. Obviously, as population grows, so too does the need for food supplies, resulting in increased pressure on agricultural resources. The way that agricultural production can be increased, at least in the short term, is by increased inputs of fertilizer, pesticides, energy (to power machines to terrace hillsides, drain swamps, etc.), and irrigation water. All of these inputs, in turn, have pollution problems which in some cases may actually result in a decrease in agricultural production.

Before discussing agricultural problems, it might first be worthwhile to examine how well the people of the world have been fed in the recent past. There are many estimates of the number who starve to death annually. Such estimates are vague because—(1) the countries where people starve often have the least adequate census systems;—(2) it is not seen as good for national pride to have starving people in a country, so often there is a perhaps intentional attempt to underestimate the problem (for example, the 1973-74 famine in Ethiopia was long denied publicly by the government);—(3) most people who are starving generally are so weakened by starvation that they succumb to some other disease.

But granted all of these stipulations, reasonable estimates are still forthcoming, and, over the past few years, ranged from 10 to 20 million people per annum. But a far greater number than this is short of food, though actual death might take a long time. Also, it is best to bear in mind that the critical shortage is not carbohydrates, but protein, and more specifically, high quality protein such as found in meat, animal byproducts, and a few seeds such as soyabeans. Starvation often occurs even though food is available, simply because there is a protein shortage. Medically, this is known as kwashiorkor disease.

Just how critical is this shortage of protein and calories? The United States President's Advisory Panel on the World Food Supply[5] in 1967 estimated that 50–60 per cent of the population of the less industrialized nations are inadequately nourished. That is, one-third of the population of the world! A 1974 United Nations report estimated that as many as 800 million people, almost a quarter of the world's population, may be suffering from malnutrition. History records more acute shortages in individual countries, but it is doubtful whether such a critical food situation has ever been so world wide.

For the future, and in spite of recent developments in agricultural technology, most estimates are of increased food shortages and mass starvation. A United Nations report has warned that up to 500 million may starve within the next decade.[6] Dr. Boerma, director general of the United Nations Food and Agriculture Organization, predicts that "the world is heading towards an alarming food crisis".[7] One author who reviews many of these predictions for a worsening population-food situation states, "Most authorities agree that at the present time at least half of the people inhabiting our planet either consume too little food or have serious imbalances in their diet ... the severity of this [food supply] problem is likely to increase during the remainder of the present century."[8]

In many respects, the starvation crisis we face epitomizes the environmental problems faced by the contemporary world. The growing population, the growing rate of starvation, the self-defeating technical fix solutions, the shortage of inputs, the deterioration of the agricultural land base, and the problems of various agricultural pollutants, all conspire to produce a true environmental crisis. It is the interrelated nature of the environmental problems and the need for a systems approach to a solution, if one exists, that is nowhere more highlighted than in the crisis situation of mass starvation.

THE ENVIRONMENTAL CRISIS:—CONCLUSION

The interrelated nature of environmental problems such as population, pollution, and resource depletion has been emphasized throughout this chapter. The interrelatedness is also discussed in chapter 4, and is implicit in every other chapter where environment is discussed. This interrelatedness allows us to refer to *the* environmental problem.

The factor of time was also referred to in several parts of this chapter. We found for example that total population is growing at such a rate that it will double in thirty-three years. The assured supply of many resources, currently thought of as essential to our developed economy has been found to be uncomfortably short (thirteen years for mercury, fifteen years for tin, and twenty years for petroleum). Rates of starvation in the world seem to be rising rapidly, putting another time factor on the environmental problems.

It is really this time factor which changes the environmental problems into environmental crisis. In so many respects, the life support systems of the earth are being threatened. While solutions for various parts of this problem are possible (chapter 3), the time factor makes such solutions less likely. Time is the most scarce of resources in the environmental problems dilemma. This is *the environmental crisis*.

3 A review of proposed solutions to the environmental crisis

In line with the widespread recognition that there exists an environmental crisis, there has been a proliferation of written material, outlining what is the *real* problem, and what is the *real* or only solution. In reading this material, several observations can be made.

While there is general acceptance of the idea that a serious environmental crisis confronts mankind, and while there is reasonable agreement as to the characteristics of this crisis, there is considerable disagreement over the relative seriousness of the different manifestations of it. Thus, for example, while Paul Ehrlich argues that uncontrolled population increase is the main or underlying factor from which all other problems flow, Barry Commoner claims that harmful technology holds this position.

There is also quite strong disagreement concerning the immediacy of the environmental crisis. This lack of agreement concerning the time available to find and implement solutions often underlies the solutions which are suggested. There is, however, fairly general agreement in much of the environmental literature that solutions do exist. But while some writers argue that mankind will surely adapt to new conditions, and implement solutions, others are quite pessimistic that mankind will make use of solutions which already exist, or which may be developed.

On the basis of these differences a quite fundamental breakdown of all popular environmental crisis solution theories into five categories can be made. The solutions represented by these classes are: (1) scientific; (2) technological; (3) economic; (4) philosophical; (5) sociological. Each of these will be discussed in turn.

1.—SCIENTIFIC SOLUTIONS

This approach sees the solution to any and all environmental problems within the realm of science. It is argued that either more scientific research or a redirection of scientific research could solve the environmental crisis.

John Platt, in talking about the environmental crisis, argues that since it is the new science and technology that have made our problems so immense and intractable, then solving these problems will require something very similar to the mobilization of scientists for solving crisis problems in wartime. "Where science and technology have expanded the [environment] problems in this way, it may be only more scientific understanding and better technology that can carry us past them. The cure for the pollution of the rivers by detergents is the use of non-polluting detergents. The cure for bad management designs is better management designs."[1]

Platt simplifies all the environmental problems and considers they are amenable to scientific research and applications, but concludes: "The task is clear. The task is huge. The time is horribly short. In the past, we have had science for intellectual pleasure, and science for the control of nature. We have had science for war. But today, the whole human experiment may hang on the question of how fast we now press the development of science for survival."[2]

Irving L. Horowitz, an American sociologist, also subscribes to the scientific school of thought when he argues, "if the [environmental] problems are scientific in nature, should not the resolutions be scientific in nature?"[3]

René Dubos argues that a new type of scientific enquiry using holistic premises must be extended to problems such as defining the limits to the range of human adaptability, to problems such as social stress, air pollution, etc. Similarly, the biological and social factors affecting population growth could be amenable to scientific study.

B.F. Skinner, a behavioural psychologist, also argues that the *usual* scientific approach will not work in solving the environmental problems. "The application of the physical and biological sciences alone will not solve our problems because the solutions lie in another field. Better contraceptives will control population only if people use them. New methods of agriculture and medicine will not help if they are not practised, and housing is a matter not only of buildings and cities but of how people live. Overcrowding can be corrected only by inducing people not to crowd, and the environment will continue to deteriorate until polluting practices are abandoned."[4]

Skinner appeals for a science of human behaviour, out of which can be developed a technology of behavioural modification. Skinner rejects the idea that man has a free will, or is capable of choice, arguing that instead his behaviour is determined totally by his environment. He argues that a science of human behaviour will "follow the path taken by physics and biology, by turning directly to the relation between behaviour and the environment and neglecting supposed mediating states of mind"[5].

In arguing the morality of scientific study and control of human behaviour, Skinner explains, "To refuse to control is to leave control not to the person himself, but to other parts of the social and non-social environment."[6] In other words, man is completely subject to control, and all Skinner wishes to do is substitute for the capricious and environmentally destructive controls now existing, the sane, scientifically developed behavioural controls which will ensure the safety and survival of mankind.

Skinner even argues that the science of behaviour, unlike other physical and biological sciences, contains within it the basis of value judgments.

He concludes that only through the open acceptance of the science of behaviour can we hope for a solution of the serious problems which confront mankind. At present, ethical squeamishness allows only influence and not control—and thus condemns behavioural science to a position where it cannot help but fail. "Our culture has produced the science and technology it needs to save itself ... But if it continues to take freedom or dignity, rather than its own survival, as its principal value, ... [it will find itself] in hell with no other consolation than the illusion that 'here at least we shall be free'."[7]

2.—TECHNOLOGICAL SOLUTIONS

The people who argue for technological solutions to any and all of the global environmental problems are many, and come from diverse academic backgrounds. In general, the technical fix type of argument has a basis of ethnocentricity, rather extreme pragmatism, and a somewhat anti-humanistic concept of mankind.

Athelstan Spilhaus, past president of the American Association for the Advancement of Science, argues that technology has the answer to the environmental problems confronting mankind. "The great challenge is how to continue providing for people's needs and wants and yet, at the same time, to manage the environment by containing wastes in the manufacturing plants—by recycling, reprocessing, and reuse—and by rebuilding industry to be saving both of materials and energy."[8]

Technological utopian thinkers like R.G. Siu talk of the golden era which confronts mankind, when technology will solve all of his problems and allow mankind to develop into an even higher species. "The next fifty years [will] provide technology the opportunity of a millenium ... The inevitability of the intrinsic logic of the systems view extends the scope of technology beyond the bare material necessities of animal life to the grandeur of art and spirit. It [technology] is now focusing attention on values with man as the measure."[9] This argument is carried much further by Berry[10] who argues that over the next ten thousand years virtually all man's problems will be solved and all man's needs will be met, thanks to the wonders of modern technology.

A much better known technological utopian thinker is R. Buckminster Fuller. He argues that most of what we regard as problems today can be solved by technology. Problems such as pollution of air and water can be easily solved by technology (he asserts), but we do not do so because it costs too much. Through using Fuller's technological utopian concepts, nothing ever costs too much, in any way, for mankind. "Wealth is our organized capability to cope effectively with the environment in sustaining our healthy regeneration and decreasing both the physical and metaphysical restrictions on the forward days of our lives."[11] "Sum totally, we find that the physical constituent of wealth—energy—cannot decrease and that the metaphysical constituent—know-how—can only increase. That is to say that every time we use our wealth it increases."[12]

Barry Commoner, a well-known American popular ecologist, also subscribes to the school of the technical fix. Commoner argues that the underlying cause of all of our environmental problems is the way our technology has changed; how polluting, resource-exploitive and energy-demanding technology has come to replace nonpolluting, resource-saving, light-energy technology. He argues that while production for most basic needs has just kept up with the increase in population, the kinds of goods produced to meet these needs have changed drastically. New production technologies have displaced old ones.

To prove this point, Commoner analyzed post-World War II American industrial production figures.[13] His findings indicate that a group of inorganic, nonreusable products had tremendous growth rates, replacing organic or reusable products, many of which declined relative to population, or in total. For example, production of synthetic fibres was up 5,980 per cent, while cotton was down 7 per cent, and wool down 42 per cent. Aluminium was up 680 per cent, while lumber was down 1 per cent. Nitrogen fertilizer (1,050 per cent), synthetic organic chemicals (950 per cent), pesticides (390 per cent), and motor fuel consumption (190 per cent), all showed startling rates of increase. Thus, the type of industrial activity which has shown the most rapid increase is also the type of industry which plays the most havoc with the environment in terms of pollution and resource demand.

The solution to these technologically caused environmental problems is to develop and use new technology which is ecologically conceived. Commoner argues that "technology properly guided by appropriate scientific knowledge can be successful in the ecosystem, if its aims are directed towards the system as a whole rather than at some apparently accessible part. Ecological survival does not mean the abandonment of technology. Rather, it requires that technology be derived from a scientific analysis that is appropriate to the natural world on which technology intrudes."[14]

3.—ECONOMIC SOLUTIONS

The people who argue that economic solutions can be found for the environmental problems tend to follow certain assumptions. They see man as being an economically rational creature, allowing economic factors to outweigh social and ethical factors in any decision-making process. The economic system is seen as being amenable to rational change, which will not unduly conflict with other parts of the social system of which the economic system is but a part. The environmental problems examined in economic solution arguments are generally pollution and resource depletion. A few theorists have approached the population problem with suggestions of taxes and subsidies, but these are almost always mentioned in a peripheral fashion, and show evidence of little serious research. The Natural Resource Unit scheme, discussed later, is an exception to this criticism. The more metaphysical aspects of the environmental problems, such as types of lifestyle, ethical approaches to nature, man's conception of his position in the natural order, etc., are rarely if ever considered by the economic solution proponents.

One of the most vocal proponents of economic solution is John Maddox,[15] past editor of *Nature*. He agrees with the doomsday prophets, as he calls them, that disaster will come if things go on as they are going now. He does not, however, believe that catastrophe can only be avoided by Draconian remedies, which, he suggests, are sought by Ehrlich and other environmental Cassandras. Maddox believes that we should be able to cope with a population double that at present, by the end of the century, providing all with both adequate food and shelter. Pollution, he believes, is only a question of price, of economics. We can reduce pollution if we are prepared to pay for it. "The issue [pollution] is a simple economic one. How much are people prepared to pay, either directly through the cost of anti-pollution programmes, or indirectly in the form of higher prices as when the polluter is made to pay?"[16] A very similar argument has been presented by the Australian economist, Colin Clark.

Dr. H.C. Coombs argues that economic changes can solve many (though perhaps not all) environmental problems. The price mechanism can be used to economize scarce resources to whatever degree is desired. Pollution could be controlled economically, by relying on "complex regulations involving supervision and the imposition of penalties, or alternatively on excise taxes on the commodities produced by the offending processes, or preferably on the polluting effluents themselves".[17]

As an overall policy to alter the present social and economic system toward ecological stability, Coombs proposes, "to modify the present system of income distribution so as to reserve to the State part of the gross national product before it is allocated as personal and corporate incomes". Similarly, in order to reduce the harmful effects of crass materialism, "a prohibition on advertising or its exclusion from acceptable costs for the purpose of taxation would appear to have attractions".[18]

By far the most comprehensive and logical economic solution theory has been devised by Dr. W. Westman and Dr. R. Gifford, two ecologists.[19] In their scheme, designed not to replace but to complement the existing money-based economy, a certain number of Natural Resource Units (NRUs) would be distributed to each person each year. For any activity which has an environmental impact, a number of NRUs would have to be surrendered, the exact number depending on the environmental impact. NRUs could not be transferred, nor could they be bought for money. They could, however, be saved up from year to year. A computerized credit card system would keep account of each person's NRU supply and spendings. The possibilities of overspending NRUs (e.g., unwanted pregnancy), the effects on one nation if the scheme was not taken up elsewhere, the impact of an NRU scheme on underdeveloped nations, and even the political possibilities for implementation of such a scheme, are all discussed and explained by Westman and Gifford.

They believe that an NRU scheme, even though involving more governmental planning and regulation than is currently deemed feasible or acceptable, would lead to less restriction of personal freedom in a steady state than would result from the current trend towards unsystematic imposition of governmental regulations.

4.—PHILOSOPHIC SOLUTIONS

This line of argument sees the problems of pollution, overpopulation, resource exploitation, etc., as being mere symptoms of far more profound problems. It is argued that the way man perceives himself in relation to other people, nature, and the ecosystem is fundamentally wrong, and leads to current environmental problems. Attempts to solve the environmental problems through science and technology will fail, because new problems, perhaps worse, will merely replace the old ones. Economic solutions likewise avoid solving the fundamental problems, so will have little permanent effect in correcting the environmental problems. It is argued that only by changing man's philosophical or ethical stance can environmental solutions be found.[20]

Lynn White, an historian, argues that the main fault in the environmental crisis is man's concept of nature as something to be exploited and mastered by man, arguing that this concept follows directly from the development of Christianity out of Judaism. "Christianity, in absolute contrast to ancient paganism and Asia's religions ... not only established a dualism of man and nature but also insisted that it is God's will that man exploit nature for his proper ends ... By destroying pagan animism, Christianity made it possible to exploit nature in a mood of indifference to the feelings of natural objects."[21] White then goes on to argue that since the underlying environmental problem is the Christian concept of man, vis-a-vis nature, only by changing this philosophical stand, will solutions be found. "What we do about ecology depends on our ideas of the man-nature relationship. More science and technology are not going to get us out of the present ecologic crisis until we find a new religion, or rethink our old one ... We shall continue to have a worsening ecologic crisis until we reject the Christian axiom that nature has no reason for existence save to serve man ... Since the roots of our trouble are so largely religious, the remedy must also be essentially religious."

Edward Fiske, a religious expert, supports the analysis of White, talking about "what may prove to be the most far-reaching new religious issue of the 1970s—the theology of ecology, or man's relationship to his environment".[22]

Henry Skolimowski[23] strongly rejects this whole line of Judeo-Christian argument, stating that it is to early scientists like Francis Bacon and Galileo Galilei, and their experimental methods, that credit or blame for concepts of man mastering nature must be accorded. Isaac Newton finalized the development of this concept of nature as there to be mastered by man.

This concept of nature as there to be exploited by man was challenged by Rousseau as the first of a long line of thinkers, through Thoreau, Emerson, the transcendentalists, and the Nimbin or Earth Garden philosophy. Skolimowski argues that this change in philosophy, from seeing nature as there to be exploited, to seeing nature as good and sacred (Mother Earth) is a trend which can perhaps solve current ecological problems. Society must work towards a new social utopia which will require the development of a religious reverence for nature.

Eric Aarons[24], editor of *Australian Left Review*, also follows the philosophical solution theory, but he tends to argue basically for man to have a new humanistic concept of himself and his fellow man. Aarons sees hope in the Consciousness III ideology as traced out by Charles Reich in *The Greening of America*. By using a concept of the dialectic, he argues, by seeing the world of man and nature in a state of continual flux, man can move towards development of a sane, ecologically sound philosophy. Australian poet and environmental activist Judith Wright McKinney supports this philosophical solution argument, seeing the need for a capacity to change individual human values. She argues that "organizations are powerless to change themselves. Only individuals can change themselves, and if enough of them do this, they in turn can change organizations."[25] The type of change required for the value perspective of mankind, she argues, is related again to the concept of man and nature. We must delete from our minds the scientific view of man being separate from nature.

Leo Orleans and Richard Suttmier take the view that the teachings of Chairman Mao could help correct our erroneous environmental philosophies. "Maoism", they state, "is first and foremost an ethic of frugality, of 'doing more with less'."[26] One of the main teachings of Chairman Mao has been that of comprehensive use of any and all resources. What is a byproduct or pollutant from one process must be made a resource for some other process.

This ethic of frugality argument has also been applied here in Australia, for example, in an article I published in a conservation journal.[27] In it I argued that resource demands, pollution, and materialism could all be decreased by a change in our philosophic stance, from one of exploitation to one of frugality. As well, I argued that a more frugal existence actually results in an improvement in the condition of man. This is, of course, again the old arguments of Rousseau, Thoreau, Emerson, & Co. There is no proof of this—only faith.

Another Australian who has written on this subject is Charles Birch, professor of biology, who argues that environmental stability would require "a revaluation in our values perhaps not radically different from that in some Scandinavian countries today. The overriding objective would be an Australia characterised not by power and bigness but by elegance and quality. There would be a new emphasis on the virtues of the small, the simple and the non-violent industries."[28] Only this change in social philosophy will allow mankind to correct environmental problems. Birch stresses the critical importance of how man perceives himself in nature. The idea that nature exists to serve mankind cannot but lead to ecologic disaster.

5.—SOCIAL CHANGE SOLUTIONS

Those who argue for social change solutions to the environmental problems generally argue that pollution, resource depletion, overpopulation, etc., are merely the symptoms of a deep social malaise. It is the way we live, the social contacts we develop, the type of family we form, and how we perceive and develop our relationships within the social order which determine, ultimately, the impact we shall have on the environment.

The environmental solution ideas discussed in this section really refer to changes in lifestyles rather than social change per se, where lifestyle is seen as the behavioural manifestations of culture. The term social change is retained, however, as the heading for this section, because that is the sort of term many of the authors under review would use. While they would talk of social change the main concern is changed social behaviour, i.e., changed lifestyle.

The ethical and philosophical aspects discussed in the past section are seen as relevant only when they are implemented in behaviour. Economic solutions are seen as being too peripheral to solve the real problems, only correcting symptoms. Similarly, the science and technology which we use, are products of our social system, and will not be changed to solve environmental problems, until the social system itself is changed. The social change solution theories then argue that only significant or radical changes in the way we live will solve environmental problems. The nuclear family, repressive sexuality, materialistic lifestyles, etc., have all been suggested as aspects of our culture which must be changed.

Barry Weisberg[29] argues that a precondition for our survival is the transformation of the way in which we live. This change in lifestyle, he argues, is also tied in with the changes needed to solve racism, poverty, and the exploitation of third world nations.

Paul Ehrlich, the American biologist who has done so much to warn the world about problems of overpopulation, argues that solutions to environmental problems must be social solutions. In *The Population Bomb* Ehrlich concludes that "there is considerable reason for believing that some extremely fundamental changes in society are going to be required in order to preserve any semblance of the world we know".[30] Such changes include changing the family system, the school system, our economic system, industry, etc.

Within this new society, we must reduce our material wants to a level which can be sustained by a stable population over a long period of time. At the same time, we must encourage the full development of creative energy drawn from our greatest natural resource—our potential as human beings. Our end must be a life of satisfaction for each individual; our means must be free self-expression of the individual compatible with the rights of all other human beings ... Too many have been told from childhood that sexual pleasure is bad. Too few realize what good food tastes like ... Crowded, smoggy cities limit the opportunities for enjoying outdoor sports, solitude, or the wonders of nature. There are, in fact, enormous intrinsic advantages and joys in simplifying our life-style."[31]

Continuing the argument for social solutions to environmental problems, Ehrlich and Harriman state that, "Young people today in many countries are already experimenting with new life styles, some of which might well be incorporated into the culture of the new man. Communal life, for example ... certainly fosters a spirit of community and co-operation among those participating ... Such a spirit of co-operation, if it can be extended to include all of humanity will be essential if we are to solve our planet wide problems.

The rewards of marriage, indeed the institution of marriage itself, will in the future be very different from our traditional one ... A variety of marriage styles ... informal, easily dissolved marriage or ... group marriage."[32]

Murry Bookchin argues that the attempt to blame environmental problems on technology (Barry Commoner) or overpopulation (Paul Ehrlich) is an attempt to divest ecology of its explosive social content. He sees that the roots of the ecological crisis lie precisely in the coercive basis of modern society. Bookchin argues that reform and piecemeal solutions to environmental problems are harmful in the long run. Only radical and complete social change will offer any hope of ecological stability. "It is necessary to overcome not only bourgeois society but also the long legacy of propertied society; the patriarchal family, the city, the state—indeed, the historic splits that separated mind from sensuousness, individual from society, town from country, work from play, man from nature. The spirit of spontaneity and diversity that permeates the ecological outlook toward the natural world must now be directed toward revolutionary change and utopian reconstruction in the social world. Propertied society, domination, hierarchy and the state, in all their forms, are utterly incompatible with the survival of the biosphere. Either ecology action is revolutionary action or it is nothing at all. Any attempt to reform a social order that by its very nature pits humanity against all the forces of life is a gross deception and serves merely as a safety valve for established institutions."[33]

With sentiments quite similar to those expressed by advocates of the Nimbin philosophy in Australia, Bookchin predicts that after the social-ecological revolution, "in this decentralized society, a new sense of tribalism, of face-to-face relations, can be expected to replace the bureaucratic institutions of propertied society and the state. The earth would be shared communally, in a new spirit of harmony between man and man and between men and nature."[34]

CONCLUSION

With a few exceptions, these numerous suggested solutions to the environmental problems are not mutually exclusive. To argue that technology will solve or alleviate an environmental problem does not mean that social change may not also be required. Thus, it is quite obvious that technology can be used to alleviate some immediate environmental problems such as air and water pollution. Technology can also be used to allow solar energy to replace petroleum and atomic energy, and the more comprehensive usage of resources (recycling). Scientific developments are required to facilitate these technical developments. Often such technology may be developed and implemented only if there is sufficient economic inducement. A change of attitude—to a recognition of quality of life being separate from, and in some respects in conflict with economic growth would be required before such economic inducements could be introduced. The result of such a change in attitude and economic priorities could lead, to some extent, to changes in the social order. How profound such changes would be is obviously debatable.

This is the very point which will be developed in chapter 12, and utilized in chapter 13. There are no simple answers. To a certain extent any and all of the proposals for environmental solutions can have some alleviating effects. The real problem comes when proponents argue that their solution is *the* solution. That what they suggest would solve our *entire* ecological crisis and that any other solution ideas are too superficial, or confuse the issues, or miss the essential point, or even make the problems worse.

The type of solution to the environmental crisis proposed by any one author depends on:

a. The author's understanding of the objective data relating to the problems.

b. The time span over which the problem and the proposed solution are seen.

c. The value system within which the author exists and within which he analyzes the problems and proposed solutions.

d. Academic or other professional reasons inducing one to take, or adhere to, a particular form of argument perhaps because it seems original, or because it contradicts someone towards whom the author is personally opposed, or perhaps because it is seen as an academic coup.

In the following chapter it will be explained how most of the previously mentioned environmental solutions have some relevance, but that the critical factor is the framework within which they are analyzed. A dynamic systems framework is introduced to facilitate an understanding of the environmental crisis and its proposed solutions.

PART 2

A model of the environmental crisis

In Part 2 of this book a model is introduced to provide a framework for analysis of the environmental crisis, and the proposed solutions to that crisis. This model is based on systems theory, and more specifically, on what is known of the workings of natural systems. The theoretical background of support is discussed in chapter 4. The model, presented graphically, is meant to be seen as a heuristic model, i.e., an aid to understanding.

In chapters 5–11, this heuristic model is discussed in depth. The intent in each of these chapters is to illustrate the nature of the inter-relationships between the elements comprising this model. To do this several specific examples are selected and discussed in each chapter. Justification of my arguments often of necessity must depend on my quoting the opinions of other authors who have worked in specific areas under discussion. A wide range of sources from newspapers to journal articles to books are cited. Since the model being developed here draws from what are traditionally several quite separate academic areas, this method is the only one applicable, in my opinion. While in each chapter the relationship between only two of the elements is discussed at any one time, it is imperative that the overall model (shown graphically at the start of each chapter) be borne in mind. It is the relationship of the elements within the overall model which is important.

4 A model for understanding environmental problems and solutions

The environmental problems outlined in chapter 2 and the various solutions proposed in chapter 3 are all normally seen in a linear, cause-effect relationship. A problem, such as air pollution, soil erosion, overpopulation or genetic mutations caused by atomic radiation, is normally outlined, then a solution to this problem is sought. There is the general assumption that it is possible to solve such a specific problem, and that in fact specific problems exist as separate entities.

In this chapter, it will be argued that the so-called problems as well as the solutions to these problems are all elements of a complex, intermeshed system, an understanding of which is necessary for an adequate identification of the problems and specification of solutions.

THE INTERRELATED NATURE OF ENVIRONMENTAL PROBLEMS

Many of the environmental problems outlined in chapter 2 are related to each other in very complex ways. For example, to speak of *over*population obviously implies some benchmark against which to compare the *over*population with the implicit *optimum* population. Such a guide could be drawn from a comparison with resources, with pollution, with food supplies and starvation, with social factors like murder, unemployment, or family change, or, simply, even with a subjective, aesthetic judgment (as, for example, ochlophobia, or the fear of crowds or of too many people). Science alone does not show that there is overpopulation.

Similarly, to speak of resources being limited is only meaningful when we clarify the size of population and rate of population growth, the rate of resource use and the rate of change in amount and type of resource use, and the price in terms of efforts and foregone consumption we are willing to make to acquire that resource. Similarly, any discussion of availability of resources and efficiency of usage, and, in fact, the need for that resource at all, presupposes a certain technology as well as a certain social system.

Gold, for example, is a scarce (and valuable) mineral, almost completely because of social factors. If gold was not seen as being inherently valuable, as a source of, or the support of, most money systems, then presumably there would be more than enough gold to satisfy the limited industrial and aesthetic needs. The perceived scarcity and value of gold are tied inextricably to our social and economic histories. As well, the technology of gold mining at any time has an impact on the actual supply.

In a similar vein, concern over food supplies only has relevance with regard to other factors. Technology exists to produce larger food crops at any given time, but the long-term effects of this technology on the soil, the pollution effects of the fertilizers and pesticides, and the salinity problems caused by irrigation all must be considered. There are also problems of what kind of food to produce and consume, and these factors, being largely socially determined, are not amenable to any technical fix.

Environmental problems occur within what can best be conceived of as a world ecosystem. This ecosystem is, as Weiss has pointed out, "a paradigm of the principle of interdependencies, partly prestructured, partly in free systemic interaction which make it possible for organisms to mesh harmoniously with their environment and with one another ..."[1]. Buechner goes on to argue how "the concept of an ecosystem as one end of a spectrum of heirarchically organized systems of increasing complexity, from atoms upward, provides an indespensible tool for understanding how nature, including man, is structured and how it works".[2]

Thus it is as a heuristic concept that ecosystems are posited by these authors, as well as in this book. All lessons from ecology point us toward such a systems analysis of environmental problems. For these reasons, in this book it is accepted that an ecosystems framework is the most productive way to try to understand the environmental problems as discussed in chapter 2.

THE INTERRELATED NATURE OF SOLUTIONS TO ENVIRONMENTAL PROBLEMS

Just as we have argued that environmental problems form part of a complex system, so too do the solutions proposed for these problems.[3] For example, to introduce a technical change to solve an environmental problem can easily have social consequences.

A fairly clear example of this concerns birth control technology as a solution to overpopulation. The pill, abortion, and even vasectomy, have all been seen as threats to society, to morality, etc. While many liberals might scoff at this, it does not alter the fact that an actual relationship almost certainly exists. For example, medical research in Brisbane[4] indicates that a vasectomy operation had an impact on the sexual activity of 70 per cent of those having the operation (66 per cent improvement and 4 per cent deterioration).

As with any such sexual technology factors as the pill and self-induced abortion by a vacuum tube or by use of the newly developed prostoglandins crystals, there will be changes in sexual and social behaviour. Of course, whether the change improves or harms morals and society depends on one's subjective assessment.

This is only one example of how any attempt to solve an environmental problem may well cause changes in other areas of life. It is impossible for technology to be radically altered without lifestyle being altered as well. The economic system, if it is altered, will engender changes in technology and in lifestyle. In other words, while environmental problems are very much interrelated, attempts to solve those problems are also interrelated. It is naive to expect that piecemeal solutions can be applied without unforeseen and often dramatic consequences.

Systems Model

Detailed study of all aspects of the environmental problems, as well as the proposed solutions, leads to the conclusion that all these elements can be understood only within a complex framework. Since the environmental factors are interrelated and since a change in any one factor will have a direct or indirect effect on every other factor, a dynamic systems concept is suggested. Further, since the proposed solutions are also interrelated, a dynamic system is again suggested. For these reasons systems theory was selected as appropriate to the heuristic model proposed in this book.

A *system* can be defined simply as a set of objects together with relationships between the objects and between their attributes.[5] Within this definition, objects are simply the parts or components of a system, and these parts are unlimited in variety, and attributes are the characteristics or properties of those objects. Perhaps a more suitable working definition for the type of dynamic systems discussed here has been given by Walter Buckley. "A complex of elements or components directly or indirectly related in a causal network, such that each component is related to at least some others in a more or less stable way within any particular period of time. The components may be relatively simple and stable, or complex and changing;

they may vary in only one or two properties or take on many different states. The interrelations between them may be mutual or unidirectional, linear, non-linear or intermittent and varying in degrees of causal efficacy or priority. The particular kinds of more or less stable interrelationships of components that become established at any time constitute the particular structure of the system at that time, thus achieving a kind of 'whole' with some degree of continuity and boundary."[6]

The development of the system proposed in this book grew out of the study of environmental problems (chapter 2), and various proposals to solve these problems (chapter 3). Because the system is meant to form a heuristic model, specification of that system must depend on heuristic relevance. A heuristic model, unlike a simulation model, does not attempt to accurately represent or mime objective reality, but rather to present an interpretation of that reality which facilitates its comprehension. For example, Freud's concept of the ego, super-ego, and the id is a heuristic model of the human mind, but there is no suggestion of a brain subdivided into three compartments! Presumably Freud chose these terms because, in his opinion, this model of the mind best promoted understanding of mental illness and health.

In the social sciences the most widely known heuristic type would be economic man, though no one would suggest that such a super-rational person exists. The test of a heuristic model is not how closely it approximates objective reality (simulation), but how much it promotes understanding, and provides a basis for decision making.

As already explained, the heuristic model suggested in this book consists of a system and its component subsystems. The selection of specific elements to form this system resulted from the extensive reading involved in Part 1, as well as considerable trial and error. Eventually I found that a system conceived of as having four subsystems was the simplest heuristic model capable of incorporating all factors involved in the environmental crisis, yet still capable of providing a framework for environmental planning.

The system proposed here has four primary elements or components, to be called *ethics*, *environment*, *lifestyle*, and *technology*. Each of these four elements in reality is a subsystem and contains within them further subsystems. The four elements or major subsystems together make up the system with which this book deals. The names chosen for these four elements or components, being popular, common terms require definitions.

The *environment* refers to the total physical situation within which the other subsystems operate. It includes natural, man-modified, and man-developed aspects.

Ethics in this case can be understood as concepts of moral conduct, as ideas of right and wrong, good and bad, proper and improper, etc.

Technology refers to methods or techniques of production, and provision of services. It includes scientific learning, machinery, medicine, and all other tools used by man to achieve desired ends.

Lifestyle, Michelson argues, "is based on role emphasis". Each person has many roles which he may fill at different times of his life, or even within one day. These many possible roles can generally be summarized into five spheres: political control, economic supply, propagation, socialization of the young, and explanation of the supernatural. "Together these elements help define the content of roles. Lifestyle, then, is a composite of these aspects of the roles a person strongly emphasizes. It refers ... to styles of living".[7] For purposes of this book, *lifestyle* can be taken as the behavioural manifestations of all that is entailed in culture.

For simplification and understanding it is easiest to conceive of the four elements or subsystems, *ethics*, *environment*, *lifestyle*, and *technology*, as being either on the face of a sphere, equidistant from each other, or for purposes of visual simplification, to picture the *environment* in the centre, with the other three elements of *ethics*, *technology*, and *lifestyle* spaced in a circle around *environment*.

        ethics   lifestyle

             environment

             technology

Since this book is concerned initially with environmental problems and solutions, environment is graphically placed in the middle, but does not, in fact, have a primary position in this model. Within the systems framework proposed here, there are information flows from each element, or subsystem, to every other element or subsystem. This information flow can vary from subtle to obvious in its cause-effect relationship. The information flow between the various elements or subsystems at any given time is more significant than the actual state of the elements. Information in this sense can be defined as that which alters the other elements in any way. Information constantly passes between the four components, and at any given time determines the nature of these components and of the overall system. It is the flow of information which is critical.

The direction and ubiquity of this information flow can be represented graphically by figure 3.

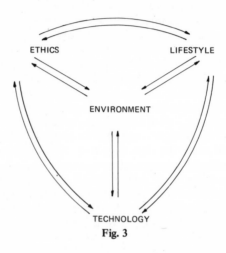

ETHICS   LIFESTYLE

ENVIRONMENT

TECHNOLOGY

Fig. 3

All this is saying is that changes in any one of the four elements or subsystems may well have effects on any or all of the other elements or subsystems. That technological innovations may affect ethics,.environment, and lifestyle. Further, that the induced change in the element or subsystem may have a further effect on a third element or back on the original element. The information flow, in fact is multi-directional, omnipresent and continuous, with the whole system in a state of change, of ebb and flow.

While the direction of and content of the information flow may be reasonably open to clarification, the meaning which this information has to another component of the system is far more complex. The meaning of the information flow, or the effect it has on other components, will depend on what other information is provided simultaneously. As in any system, there are greater or lesser tendencies toward dynamic equilibrium or disequilibrium, not only of the whole system, but within each subsystem as well. The tendencies towards equilibrium and optimization in the subsystems may well not be in accordance with equilibrium and optimization of the complete system. The system may then impose constraints on to the subsystems to facilitate system equilibrium and optimization.[8]

An example of how this dynamic system works can be drawn from the vasectomy example discussed earlier in this chapter. To the extent that vasectomy increases the male opportunity to practise contraception, and allows intercourse free from the fear of unwanted pregnancy, it obviously does have some effect on moral values or ethics, as well as altering lifestyles. Again, whether this is good or bad is a subjective assessment, but the impact on ethics exists, nevertheless.

Vasectomy, as an example of medical technology, also has an environmental influence. There is a direct influence of fewer children (since that is the purpose of vasectomy). There would also exist indirect influences such as reduced need for pills and other birth control devices.

But obviously none of these effects of vasectomy end there. A lower birth rate, because of vasectomy, will further reduce demands for resources, will alter the age composition of the population, and will alter other social factors, over time, such as the perceived optimum family size. To the extent that vasectomy reduces the actual rate of increase in size of population, it will have effects on social factors such as crime rates, juvenile delinquency, and even suicide.

That such flow on effects are vague and hard to pinpoint clearly does not detract from their existence and effectiveness. Obviously, any one of these tenuous influences may not have the suggested effect, because it is offset by some other influence. That is the nature of a dynamic system.

Each of the four elements or subsystems, ethics, lifestyle, environment, and technology, also experience information flows from within. That is, each element as a subsystem is in itself in a state of change, with the usual tendencies towards stability and instability.

To complete the graphic presentation of this concept, the overall model may be seen in figure 4.

This model is merely a development showing graphically that each of the four elements or major subsystems is subject to influences not dependent on the other elements or major subsystems. That is, that the environment is subject to changes which result from only environmental causes, or that technological factors may well induce technology change, quite independent of external factors. It may well be possible, however, to trace this factor back to some influence outside the subsystem.

To carry on with the vasectomy example, the medical technology of cutting the vas deferens and tying back each half so that sperm cannot pass, is a reasonably permanent operation. Medical technology is being developed to make the vasectomy operation reversible, and thus, perhaps, more popular. This technology is a

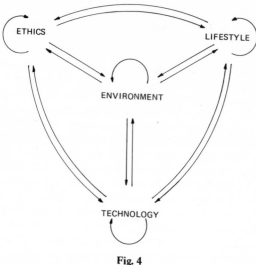

Fig. 4

development from other technological research in plastics and in techniques of arterial repair. We may step back further and find that the motivation for such development stems largely from social fears linking sterility with castration and impotency in males.

Vasectomy, as an example of medical technology, is used here in a merely illustrative way to introduce the systems model and is not meant to be construed as particularly significant per se to the specific model developed herein.

CONCLUSION

It can be found that any development like vasectomy which is seen as being related to environmental problems can be best understood within the framework of the proposed model. Such a factor can be traced in either direction, that is, as having certain effects on other elements, as well as stemming from certain other factors.

In the above diagram all the subsystems would appear to exert equal influence on the state of the system at any given time. Throughout the following chapters however it will be seen that some factors are more important than others. The technological subsystem, it will be contended, contributes most to overall system instability. Thus while all subsystems experience changes from within, as well as interacting with other subsystems, it will be found that the technological subsystem is the most dynamic, and influential. That was Toffler's conclusion in *Future Shock*. This point will be developed further in chapter 11 in particular.

In the following chapters, each of the relationships between the four given elements or subsystems is developed in some detail. As well, the equilibrating and disequilibrating tendencies of each subsystem will also be discussed.

While it is not contended that the beforementioned model is the only model to help understand these environmental problems and solutions, it does seem, after much searching, to be the most useful. It is as a heuristic tool to facilitate understanding that this model is presented. The various types of bold, assertive, clearcut statements of environmental solutions presented in chapter 3 can be understood, within the framework of this model, as part of an answer, but only part of the complex web of entwined interactions. The sort of simplistic arguments and ideas from chapter 3 may well have a valid use as instruments for public education or for exerting specific pressures on specific aspects of industry, government, or even the consciences of people. But, for profound insights into the reality of the environmental problems, a dynamic systems model, as outlined within this chapter, is required.

5 Relationship between lifestyle and ethical subsystems

Mores and *folkways* refer to some of the means by which behaviour in any society is controlled or directed. But while an outside observer might see a folkway, the person within the culture would merely know that one type of behaviour is right and other is wrong. The relationship between behaviour and concepts of right and wrong is complex, and pervades what is called *culture*. The lifestyle (i.e., behavioural manifestations of a culture) of a people, and the predominantly held ethical beliefs of that people, are intimately related. In its simplest form, to some extent, we live as we think we should, and we think that way because of how we live.

In this chapter, the relationships between ethics and lifestyle will be explored. This relationship is one of the most obvious relationships within the heuristic model being described here. Yet, in spite of this relationship being obvious, when examined, it is surprising how frequently the environmental solutions literature ignores the implications of such a relationship. As indicated in chapter 3, various writers will argue the need for changes in ethics or lifestyle, at least partially ignoring the intimate interrelationships involved.

Several fairly simple and straightforward examples of the relationship between ethics and lifestyle should help clarify what is meant here by the dynamic interrelations of these two subsystems, within the system which forms the proposed heuristic model. (Figure 5)

ETHICAL DETERMINATION OF LIFESTYLE

Man has been described as a moral animal. Behaviourists can posit various biological and psychological determinants of man's behaviour, but there always seems to remain a residual segment only to be explained as what is right, or normal, or good. This is the area of lifestyle which is directly determined by ethics.

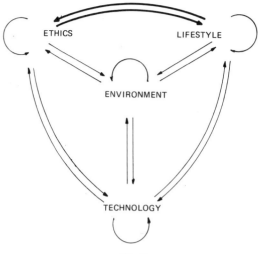

Fig. 5

The point that ethics at least partially determines the lifestyle seems so obvious that it hardly needs expansion. But the term *lifestyle*, as it is generally used, implies that there are several diverse lifestyles within one overall culture. The term *subculture*, as commonly used, has much the same intent. But while there are obviously many ways by which one can distinguish the lifestyle of an Australian student from that of an Australian farmer, and while it would be correct to point out a diversity of lifestyles led by various students, the fact remains that there are overall constants within any culture. It is usual in literature on lifestyles to point out differences, or how a deviant lifestyle differs from the norm. But this must not blind us to certain relative constants which predominate within the lifestyles of many people. And these constants are largely determined by the ethics or social norms under which the lifestyle exists.

One of the most obvious and all-pervasive aspects of our lifestyle which is largely ethically determined concerns marriage. Marriage is popular, meaning that most people at some time do marry. While there may have been other reasons for marriage in the past, if you question most people today on why they marry, it generally comes back to ethics, "it is the right thing to do".

Out of all the many possible types of family relationships, the legally constituted, nuclear family has been the only one which seemed ethically acceptable to most people in our culture. This puts very definite limits on the choice of lifestyle available to most people. And the effects flow on much further, of course, than the immediate lifestyles of the people involved. The number of children born, how they are socialized, the type of housing required, the size and type of transport, and even the directions followed by technology and industry, can all be traced back to the ethical beliefs we hold concerning marriage and the family.

Child-rearing techniques, family size, and the type of schooling we receive are also aspects of lifestyle which are largely determined by ethical concepts. The latter point has been well argued by Ivan Illich. He examines how the ethical belief in education is so deep and has such profound effects. "Today, faith in education animates a new world religion. The religious nature of education is barely perceived because belief in it is ecumenical ... Marxists, and capitalists, the leaders of poor countries and of superpowers, rabbis, atheists and priests share this belief. Their fundamental dogma is that a process called 'education' can increase the value of a human being, that it results in the creation of human capital and that it will lead to a better life".[1]

Because this ethical belief in school and education is so strong, and so pervasive, people go through much of their early lives within an imposed institutional structure which radically affects their lifestyle. Only children can be taught in school, Illich argues,[2] because adults would not submit to such authoritarian nonsense.

With regard to schooling, as in so many areas of our life, we live a certain way because of a set of ethical beliefs. These beliefs are often so entrenched and so widely accepted that we cannot see them as beliefs, but regard them as aspects of human nature, or just common sense.

LIFESTYLE DETERMINATION OF ETHICS

There are two predominant streams of thought concerning the determination of ethics. In the first (classical school), it is assumed that ethics are determined quite independently of the social system. The source may be seen as God, as natural law, or as some immutable, timeless quality which exists whether or not man is aware of it. With this sort of reasoning, man is seen as needing to learn what is ethical, and then to adjust his behaviour accordingly. But the fact that social behaviour has varied so markedly over time simply indicates how poor people are at understanding ethics, or even how they have been led astray. But the ethics always were, and always shall be the same.

The second school of thought tends to have a more recent history. Moral relativism or situation ethics it may be called. In its pure form, it assumes there are no absolute ethical values and that only within a given cultural context can something be described as being ethical or unethical. "The morality of an act is a function of the state of the system at the time the act is performed."[3] Ethics are seen here as guidelines to the smooth functioning of a social system. Problems arise when an ethical belief which was once functional becomes dysfunctional. For example, many environmental writers argue that the high value attached to having children is just such a value which, because of overpopulation, is now dysfunctional and, therefore, is unethical or immoral.

As in most such polar arguments, some sort of middle path may well be the best suited to reality. It seems from common observation, as well as a great deal of sociological writings, that concepts of right and wrong, good and bad, and of ethics, do change over time, and that the direction of this change is at least partially determined by current social reality, i.e., current lifestyle. There may well be problems of time lag, whereby social reality may well change, while popularly espoused ethics remain static. But that lifestyle does, at least partially, determine ethics, seems quite evident.

One interesting current example of this process of ethics being determined by lifestyle concerns the division of a person's time between work and leisure, and the fundamental ethical connotations of work and leisure. This fundamental split is perhaps the most significant and certainly one of the most obvious factors in the lifestyle of an individual or of a society. While it is true that ethical beliefs help determine the hours spent on work and leisure, there are numerous other determining factors, such as the type of political and economic systems within the culture. To a large extent, the ethical ideas surrounding work and leisure may be a result of the actual behaviour, and serve largely to justify that behaviour.

We now find ourselves being able to work shorter hours. During the past hundred years, we have moved from a typical seventy-hour week to a forty-hour week. But because of a strong work ethic, it seems that many people refuse to accept this increased leisure, and would prefer more money by working the same hours.[4] In fact many people indicate they would continue to work, even if there was no financial need,[5] but this choice is becoming less available. Many economists argue the environmental need for a steady state economy. In such an economy there would be less demand for labour and, presumably, more leisure. But this, then, is called the problem of leisure. Quite a number of futurological writings have noted that man's work ethic must change in order to agree with new lifestyle realities.

The philosopher, Arthur Toynbee, for example, paints a dreadful picture of the age of leisure. "The destiny of the great majority of this planet's doubled or trebled population might be to live unemployed in shanty-towns, subsisting on an inadequate dole which would be given to them grudgingly by the productive minority, who would themselves be living in fear of being massacred by the resentful unemployed majority ... it seems to me that it would be the minority [who work] that would be largely exterminated ... the majority would soon be reduced in numbers drastically by the famine and disease and mutual slaughter, and then mankind would find itself back in the state which it left behind at the dawn of the Upper Paleolithic Age."[6]

To avoid this dreadful fate, Toynbee recommends: "The unemployed majority will have to be encouraged to find satisfaction in some non-economic employment. In other words, they will have to be educated in the proper use of leisure."[7] In a similar vein, Gabor talks about the need to "educate citizens for the age of leisure".[8]

But what is really meant by this education for leisure is that people must alter their ideas and ethical concepts with regard to leisure. The lifestyle is changing to one where it is no longer possible to satisfy the work ethic by either constructive work or by "make work" work. It is therefore necessary to alter the ethic in regard to work to comply with changes in lifestyle. As long as the work ethic holds sway in the minds of men, then the reduced availability of work will be seen as the problem of leisure rather than the opportunity for leisure. This is then an example of how changing lifestyles will alter ethics.

CONCLUSION

In this chapter, the relationship between lifestyle and ethics has been explored. These two elements, as subsystems within the overall system, have the most intimate, involved, and pervasive relationships of any of the subsystems. Both are inseparable aspects of the culture with which we are involved. Yet because the various solution proposals discussed in chapter 3 do assume such a division, it is important that this model, as a heuristic model, accept this division and then proceed to point out the sort of interrelationships involved.

It has been shown that the lifestyle of a person or group is at least partially determined by ethics, and that the ethics to some extent are a direct result of the current lifestyle, or at least of the current idealized lifestyle.

Though there may often be sharp differences between how people live, and how they think they should live (lifestyle and ethics), the tendency will always be for these two factors to move towards agreement. That is, within the dynamic systems framework proposed, the lifestyle and ethical subsystems will tend towards agreement, or dynamic equilibrium. Of course, such a state (the good life) may rarely be achieved because of intervention by other factors. Nevertheless, the tendency of the lifestyle and ethical subsystems is towards a state of dynamic equilibrium.

6 Relationship between lifestyle and technological subsystems

Even at a brief glance it is obvious that relationships exist between technology and how we live. After all, we use technology to produce goods and services to facilitate our daily living in a fashion to which we have become culturally accustomed. Similarly, it does not take much imagination to see how we also adjust or adapt our lives to some of the demands of technology. Our adherence to mechanical time rather than biological time is one simple example. But perhaps because the relationship between technology and lifestyle is accepted as rather obvious, it is also seen as rather insignificant. It may, in fact, come to be almost ignored in seeking to understand social change either in traditional or modern society.

This frequent ignoring of technological factors in social change, an antimaterialist bias, can be observed in many sociological studies of social change. For example, a 1974 book of readings entitled *Social Change in Australia* does not contain one article, amongst the forty-six, which deals specifically with technology as a factor in contemporary social change in Australia. The term technology, in fact, is not even listed in the index! Exactly the same criticism may be levelled against the book of readings entitled *Australian Social Issues of the 70s* which contains not even an index reference to technology as a social issue.

As with the other five chapters of the body of this book, explaining and discussing interrelationships of the various subsystems, it is imperative that the overall heuristic model be constantly borne in mind. For the sake of clarity, the overall system is here being discussed in terms of each of its components. But understanding these components and their interactions is only possible within the framework of the overall system. Simple cause-effect relationships may seem to be suggested because of the inherently fragmentary nature of the discussion. But it is the essence of this book, that the relationships are not simple cause-effect, but of a more complex nature. While there are various relationships between lifestyle and technology, and while these relationships are to varying degrees of a deterministic nature, there are always other factors involved as well.

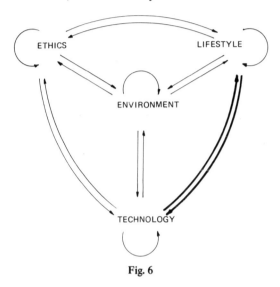

Fig. 6

TECHNOLOGICAL DETERMINATION OF LIFESTYLE

Historically, perhaps due to some innate human vanity, great social changes have generally been accredited to conscious human efforts, such as the policies or actions of a church or king. Technological change, because it seemed small and insignificant, was often ignored.

Some historians, however, attempt to redress this imbalance by pointing out profound changes in the lifestyles of people, brought about by what, in retrospect, appear to be rather insignificant technological innovations. Lynn White, for example, argues that development of the stirrup was a direct causal factor in the formation of feudal society, with all the massive changes in the lives of people therein entailed. "Few inventions", White argues, "have been so simple as the stirrup, but few have had so catalytic an influence on history. The requirements of the new form of warfare which it made possible found expression in a new form of western European society dominated by an aristocracy of warriors endowed with land so that they could fight in a new and highly specialized way."[1]

In a similar vein, White argues how invention of the iron plough, iron horse shoe, and the horse collar had profound impacts on the daily lives not only of the peasants, but of all people. The type of economic system, the size of communities, where people lived in relation to where they worked, their nutrition and health and other measures of actual standards of living were all changed because of these technological developments.

Clear-cut examples of how technology can influence lifestyle are found wherever a radical change in technology is introduced to a culture. This can take the form of a simple tool or a complicated process, but wherever the introduction is sudden and/or dramatic, the social changes are equally dramatic. The Industrial Revolution of course was an example of changed technology, altering lifestyles profoundly.

Sharp[2] has described some of the profound changes in lifestyle of the Yir Yorant Aborigines when steel axes replaced stone axes. Stone axes were always scarce, and were always owned by the men of the tribe. A woman or youth who wished to use an axe, and axes were essential for many daily tasks, had to go to a male to ask to borrow it. And the necessary and constant borrowing of axes from older men by women and children was in accordance with kinship patterns. The stone axe stood for and emphasized the superiority and rightful dominance of the male, as well as the prestige of age.

The indiscriminate introduction of steel axes via the missions had little effect on the physical standards of living. But the lifestyle was changed in other ways. Because it became possible for youths and even women to earn an axe by working on the mission, axe ownership no longer was a reserve of the older males. This breakdown of a status system led to a revolutionary confusion of sex, age, and kinship roles, with a major gain in independence for those who had been subordinated under the old system. The result of all these technologically induced changes was the erection of "a mental and moral void which foreshadowed the collapse and destruction of all Yir Yorant culture, if not, indeed, the extinction of the biological group itself".[3]

A somewhat similar situation, where Western technology has been suddenly introduced to a non-Western culture, concerns the Green Revolution, a term which refers to agricultural technology using newly developed hybrid grains, relatively large inputs of commercial, inorganic fertilizers, and the utilization of irrigation in formerly inaccessible locations. One scientist who was involved in the development of this new agricultural technology has stated: "These new varieties may be to the agricultural revolution in the poor countries what the steam engine was to the industrial revolution in Europe. Once farmers have broken with the past in agriculture, they are more susceptible to new ideas in other areas such as education and family planning."[4]

At the Second World Food Congress sponsored by the Food and Agricultural Organisation[5] delegates heard of many severe social problems brought on by the Green Revolution. These ranged from peasants being pushed off their land as land prices rose, to severe economic hardships and disparities, as only those with some capital could afford the new technology, to massive unemployment and unrest in overcrowded urban areas. Some of the radical changes in lifestyle of people, brought about by the Green Revolution, were summarized by Ehrlich. "Prices of farm land [in India] have been greatly increased by competition among landholders eager to take advantage of subsidies, while the landless rural population, already over 50 million people, is being squeezed out. The peasants are in some areas abandoning the land and moving to the already overcrowded cities. Resentment of the big landholders by the landless in India led to massive landgrabs in 1970 ... There is a growing trend toward both rural and urban violence in India."[6]

These examples indicate how agricultural technology was radically changed—and for the best of motives, yet the ensuing social unrest and resulting changes in the lives of the people may produce in the end quite different results. Social changes were perhaps not adequately allowed for in the F.A.O. programme to push the new technology. In some ways, the ensuing social changes may be counterproductive.

Ivan Illich,[7] in looking more generally at problems of Western technology introduced to underdeveloped cultures, states that "Rich nations now benevolently impose a strait jacket of traffic jams, hospital confinements and classrooms on the poor nations, and by international agreement call this development." In fact, Illich argues, technological attempts at development may well have the result of forever condemning people to underdevelopment.

It is, of course, obvious that these same sorts of relationships between technology and lifestyle exist in Australia today. For example, the introduction of sugar cane harvesters has had a profound effect on the lifestyles of numerous Queenslanders. The cane farmers have moved from being large-scale employers of transient labour, with all that that implies in terms of almost a feudal relationship, to being more of a capitalist and a machine operator. Many of the transient workers found their gypsy-like lifestyle, with its annual trek up and down the coastal route, economically nonviable. There have had to be shifts into other, more sedentary, occupations. Another group of people who have been markedly affected are the small crop farmers of Brisbane's near north coast areas. These farmers depended on the same transient workers for help at harvest time. The virtual disappearance of these workers has led to decreases in the size of small crop farms and to more small crop farmers, each of whom must do more of the labouring work himself. He is changed from a manager to a farmer. This latter trend, I believe, was also promoted by the back-to-the-land exodus of many urban Australians.

This sort of technologically induced change in lifestyle can be seen in terms of specific examples, but also as part of a long-term trend. It has been argued by Margaret Mead and Alvin Toffler, for example, that the nuclear family developed out of the extended family largely as a result of the need for increased worker mobility in our technological society, since extended families were too cumbersome. Also, a man, when stripped of the security of an extended family, is more compliant to the demands of the technological order, simply because of fear and dependence.

The needs of our technocratic society for a highly mobile workforce, devoid of multiple relationship ties, is in fact one of the most common death knells for urban communal experiments. It is one thing to arrange a move involving two people, and quite another to move six or ten, all of whom have careers. In fact, such a move normally requires all but one person to give up his career and start again, but this rarely happens. More common is the death of the communal family.

Not only does technological change on a macro level create pressures against the extended family and in favour of the nuclear family, but it may even come to outmode the stable nuclear family. Mead[8] and Toffler[9] argue that even the nuclear family may be outdated by technological developments. People of the future, they argue, will carry the streamlining process a step further by remaining childless, cutting the family down to its most elemental components, a man and a woman.

But even the isolated couple may only come to have a temporary relationship. The concept of serial marriages only of such duration as is mutually agreeable is becoming more of an accepted norm. Already, nearly one out of every four bridegrooms in America has been to the altar before. This leads to profound changes in the lifestyles of most people. Such changes, to more transitory relationships, are already apparent. Technological developments and the rapidly changing techno-industrial system, are radically changing our personal lives, even in aspects as fundamental as our type of family.

Marshall McLuhan[10] sees technologically induced changes in lifestyle being most pronounced in areas of sexual attitudes and behaviour. These lifestyle changes are, in fact, so profound that sexual concepts, practices, and ideals already are being altered almost beyond recognition. While there are numerous pressures towards the development of these radical changes in lifestyle, one of the prime factors, according to McLuhan, is the pill. By this technology, old sexual barriers and boundaries are bridged. The pill allows sexual woman to act with freedoms traditionally reserved for sexual man. This, he argues, forces a redefinition, and perhaps eventual rejection of even concepts like woman and man.

While the technologically induced lifestyle changes of McLuhan or Mead may appeal to our social conscience as good, other futurological writings paint a more mottled picture. The future lifestyle predictions of Buckminster Fuller, for example, or of B.F. Skinner may have less appeal, or may be positively frightening.

Skinner[11], devises and predicts a technology of behaviour, by means of which all human behaviour would be subject to rational control by the behavioural scientist. Any concept of freedom, Skinner asserts, is a myth in any case. This new technology of behaviour will be able to solve the fundamental problems which have for so long plagued mankind. Our irrational, erratic behaviour, our mistakes and human foibles will all be eliminated. By using the technology of behavioural modification, our lifestyles will be changed radically yet rationally, for the benefit of ourselves and of all mankind.

Lewis Yablonsky sees the control under which most people already exist as a direct product of the technological society. He calls this machine-like control *robopathology*. This technologically induced control applies to all areas of our lifestyle, from how we relate to others to how we love, to how we fight wars or produce goods. "Modern man's society, the technocratic state, and their machine domination of people have grossly altered the pattern and styles of human interaction. Their ultimate impact has been the dehumanization of people to the point where much of their social interaction is machine-like ... A social machine is a dehumanized interaction system wherein people's relationships are relatively devoid of sincere emotions, creativity, and compassion."[12]

The pernicious changes in our daily lives brought on by technology are very difficult to counter. Yablonsky argues that we must strive in two ways; to control technology, and to promote the humanistic aspects of people and groups. But even given our best efforts, Yablonsky fears that the highly structured, insidious effects of technology may continue to dominate our lifestyles producing ever more antihumanistic, machine-like behaviour. Ultimately, he sees robopathic behaviour as a threat to our very survival.

But there are also people who understand the sort of control factors which Skinner seeks and Yablonsky fears, but who are unwilling to accept their inevitability. These are people who see in uncontrolled technology an evil influence in their lives. They would wish to control their own destiny, rather than leave it to either erratic or planned control by technology. People who feel this way, and take steps to remove themselves from the system, often become what is known as the counterculture. Both Reich[13] and Roszak[14] make the point that the counterculture movement is to a large extent a reaction to what is seen as dehumanizing technology. It is of prime importance in the adopted counterculture lifestyle that technology be under the conscious control of the people, so that they can overcome the separation from their natural environment and from their selves—their alienation.

Thus we have come full circle in our analysis of technological influences on lifestyle. We started by looking at ways in which simple technological changes had rather profound effects on primitive cultures. We saw how more complex technological changes continued to result in changes in lifestyle. This led us to predictions of lifestyle becoming a direct, actively determined result of technological methods. Finally, we see a reaction against this whole sequence, i.e., changes in lifestyle, which in several respects move closer to the primitive or tribal lifestyles. This reaction or counterculture movement, to the extent that it is a reaction to technology, is, of course, determined by that technology, even though the result be the direct opposite to that intended.

LIFESTYLE DETERMINATION OF TECHNOLOGY

That technology should be determined at least partially by lifestyle seems eminently logical. What is technology but the means by which mankind sets out to achieve social ends? In fact, for technology to be determined by other than the current lifestyle seems almost an aberration since technology is totally man-made.

But while such a simplistic view seems logical, there are a number of flaws. Firstly, there are ethical values which partially determine both the lifestyle and the technology and there is no a priori reason why these two need be in complete agreement. It is quite conceivable, for example, that an ethical system which stressed both the sanctity and inherent worth of the family, as well as stressing a form of work ethic whereby production has a morally good connotation, may well lead to forms of technology which, over time, may alter or even destroy the original concepts of family, in the minds of those who developed the technology.

Secondly, no society is an homogeneous mass, either in terms of ethics or lifestyle. Since we have a wide diversity of power held by people, it is quite feasible that those with the greatest power may be able to direct technology along lines to satisfy their ethics and lifestyle, but to the detriment of most other members of the community. This charge is frequently laid against the use of cars and freeways as a technology of transport, in opposition to free mass transit systems.

Thirdly, there are ways in which lifestyle, and particularly technology, evolve or change over time, with a self-contained momentum, quite devoid of external causal factors. This will be discussed further in chapter 11. While a form of technology may well be developed to serve social ends, this may spawn other technological developments which only serve the lifestyle of a few and may even be contrary to the needs of most people.

Fourthly, while technology may be devised, or evolve, ostensibly to serve some social need, the technology may over time be found to have other, and quite harmful, effects, and may well be counterproductive of the actual ends of the lifestyle which it was meant to serve. It can be argued, for instance, that certain inorganic pesticides, if not the whole technology of chemical insect control in agriculture, have become counterproductive of the stated ends of production of good cheap food. It has been a not uncommon experience to find that when certain pesticides are used in food production, pest problems actually increase over time, production costs rise markedly, and yields may even decline. Similarly, irrigation as a technology of increasing food production may well lead not only to decreases in production, but even to destruction of the soil by salination. A UNESCO source has estimated that for every hectare of land reclaimed by irrigation in the arid zones, another hectare is lost because the soil has been sterilized by salt (salination). Research by agricultural scientists indicates that this technological backlash is now operable in southeast Queensland as well as other parts of Australia,[15] such as the Murray Valley.

Another example of a technology which was counterproductive would be the drug thalidomide which was developed as a part of a medical technology to promote health, but resulted in the destruction of health. There are a range of such examples of technology which is counterproductive of the very social aims which it was intended to serve. In some cases, as with thalidomide, the faulty technology is soon stopped. In other cases, as with certain pesticides and irrigation practices, the faulty technology may well continue to destroy, because of built-in time lags in decision-making processes. But all of these examples are exceptions. The general trend quite clearly is that technology does in some degree serve the demands of lifestyle.

Historically, there is ample evidence of technology being developed along divergent lines to suit divergent social ends. Thus the technology of the peaceful and hedonistic Minoans was used to devise home comforts such as internal lighting of buildings, and even flush toilets in 1400 B.C. The Carthagineans and the people of Syracuse, on the other hand, excelled in the technology of war, inventing torsional throwers for arrows and rocks.[16] As the Roman civilization aged, the direction of technological development shifted from military to social. Massive public baths, heated by a complex system of ducts, theatres with refined acoustics, and toys such as automated puppets[17] were all technological developments to satisfy the lifestyle of a rich, self-indulgent ruling class.

More recently, during the periods of rapid colonization by Europeans, a vast range of new technologies was developed to suit the new, by necessity self-sufficient, lifestyle of the frontier. This was a technology which utilized only locally available materials, like logs in eastern America, sod in the west, and wattle and daub in Australia, to build houses. A technology where paper greased with pig fat served as windows, where leather served as hinges, and everything from clothing to the dye to colour clothes and the soap to wash them was made at home. Such technology had to be of a small scale, simple, and requiring no, or few, imported materials. This is a form of technology which eminently suited a widely dispersed population with poor communications. This technology has largely been lost, or is only remembered in museums and books. After communications improved, settlement became more dense, and the economy changed from one of self-sufficiency to more mass production, i.e., the lifestyle altered, and the technology was altered as well.

The same sort of lifestyle influences on technology can be observed in modern society in regard to the areas where technological advances have been minor, as well as where such changes have been great. One area where technology has made notably few changes is in housing. We still use much the same materials, tools and concepts now as were used hundreds of years ago. There are minor changes and modifications but, in general, housing technology has remained fairly constant.

Just why we have chosen to largely ignore technology in terms of providing shelter is an interesting question. Toffler argues that as so many things change around us, we ensure that other things stay the same, to provide some continuity in our lives. Perhaps this is why there appears to be an ever-increasing value placed on older housing styles such as terrace houses, wide verandahs and tree-lined streets?

For whatever reasons, housing has remained expensive in terms of resources, both physical and human, and of a quality below that which technology could provide. Fuller[18] argues that by the use of technology to produce such structures as geodesic domes, cheap shelter, far superior to what we have today, could be provided for all; that if we applied technology to housing, as we have applied it to many other areas of production, there would never need be a shortage of first-rate accommodation anywhere in the world.

But, of course, we do not follow Fuller's utopian concepts of living in "a super camping structure consisting of a 600 pound, 50 foot diameter hemisphere". The reason most people still prefer a standard type of house may well be because their lifestyle would be seen as threatened by anything as new and radical as such housing technology suggests.

It is significant that while Fuller's concept of the geodesic dome for housing was first appreciated by United States military forces, the whole concept never caught on with the establishment. From being a product of military-industrial technology, geodesic domes have now become a symbol of the counterculture. Wherever the counterculture meets and settles, whether at Nimbin, Maleny, or on the bad-lands of New Mexico, the geodesic dome will be found. Such technology is only recognized as suitable accommodation by those who have changed their lifestyle significantly.

It is interesting to note that the new counterculture lifestyle, in other ways as well, utilizes a quite different repertoire of technology than is utilized in the establishment. While futuristic geodesic domes are acceptable, old-fashioned technology alone will do in other areas. Natural pest control, organic fertilizers, homecrafts such as weaving and pottery, and traditional methods of food preparation and natural childbirth are all rated highly. In fact, there is a regular quest within certain counterculture circles to rediscover the technology of several generations ago. There are also quite arbitrary rules, and great confusion over what is and is not technology which should be acceptable to the counterculture lifestyle. I witnessed a humorous example of this on a rural commune which I visited. An agricultural adviser had suggested the use of superphosphate to improve the soil and to increase their vegetable yields. One member objected, stating that such modern agricultural technology using chemicals was not acceptable. Most people agreed. But another communalist countered that, as superphosphate was only "birdshit" (guano) it was organic, and therefore within the range of technology quite acceptable to their lifestyle. The dispute was still unresolved when I left.

There are, of course, many other ways in which the shaping of technology by lifestyle could be examined. For example, largely because of changes in technology, working hours have recently been decreased somewhat. Part of this extra time is now devoted to leisure activities. One of the most rapidly expanding areas of technological development is in what could be termed leisure technology, which often amounts to an attempt to fill in time. The direction of development of this leisure technology is largely determined by the lifestyle, and in turn, of course, helps determine the lifestyle. Radio and television are examples of such technology. The use of stationary, orbiting satellites to transmit television signals around the globe is a further development, growing out of a lifestyle with high consumption needs for instant news, sports, and other entertainment. This instant, world-wide communications technology using the "cool" television, to quote Marshall McLuhan, is consistent with the changes in our lifestyle, away from superficiality, and towards more involvement. The war in Vietnam is brought into our living rooms and into our lives, and we participate in a real sense. This technological development, in turn, reshapes our lifestyle. The communication technology itself, quite separate from whatever communication is conveyed, has an impact. This is what McLuhan means when he asserts that "the medium is the message". The technology is a product of a certain lifestyle, and tends, in turn to create or shape a certain lifestyle. The global village with a tribal sense of identity is specifically the direction in which McLuhan sees our lives changing thanks to technology.

CONCLUSION

All the examples used in this chapter have pointed out a few of the myriad ways in which technology helps determine lifestyle and lifestyle helps determine technology. Historically, we saw how apparently simple technological developments such as the stirrup or the horse collar had a profound effect on medieval lifestyles. A somewhat analogous example of the steel axes altering an Aboriginal group's lifestyle was also provided. Current examples ranging from the Green Revolution to birth control were used to show how today our lifestyle is still markedly altered due to technological changes. Numerous examples, both historical and current, were also provided to show ways in which lifestyle determines the technology at any point in time. But all through the examples used in this chapter, it is important to bear in mind that the relationships are far from simple and direct, since there are many intermediary factors. Such intermediary factors are discussed in other parts of this book, and it is these intermediary factors which comprise the other subsystems.

The intent of this chapter is to explain an interrelationship within a heuristic model; only within this model as a whole can the relationship of the parts be understood. It is the relationships, the dynamic condition, which are important, not the existing state at any one time. As with the other relationships of the parts discussed here, there are tendencies towards equilibrium or agreement of technology and lifestyle. This, of course, tends to be true by definition. Changing technology becomes part of lifestyle, thus changing the lifestyle, and similarly, a lifestyle presents certain needs and wants, the satisfaction of which involves what is called technology. Only by considering both technology and lifestyle as process rather than in a static way, as in a constant state of flux, can the relationships, and the overall heuristic model, be understood.

7 Relationship between lifestyle and environmental subsystems

It is often argued that culture can be understood mainly with reference to cultural factors, as is one underlying assumption of much sociological writing. But anthropological and human ecology studies have also shown clearly that no culture is wholly intelligible without reference to the noncultural or environmental factors with which it is in relation and which condition it.

More specifically, there exists a number of intricate and intimate relationships between the environment within which a culture exists, and the behavioural manifestations (lifestyle) of that culture. The environment helps determine the lifestyle, and the lifestyle, in turn, helps determine the environment.

With regard to man's relationship with his environment, Rapoport states that: "The environment is not something 'out there'. It is not a picture or photograph admired at leisure. Man is *in* and *of* the environment. The insight of the ecological point of view is that we cannot consider either the organism or the setting by itself. Only the systemic interaction and interplay of the two is meaningful and relevant. More than that, the interaction is dynamic and active. Man structures the environment and behaves in it while searching for cues to behaviour in the environment."[1]

This dynamic, active relationship between lifestyle and the environment within which that lifestyle is enjoyed is the subject of this chapter.

ENVIRONMENTAL DETERMINATION OF LIFESTYLE

There have been various divergent views of the extent and pervasiveness of environmental determination of culture, and in turn of the behavioural manifestation of the culture—i.e., lifestyle. In the early part of this century a form of environmental determinism was

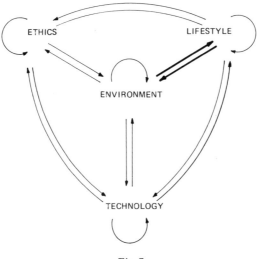

Fig. 7

accepted which stressed the effects of such variables as climate, terrain, and even soil on the regional and national character, lifestyle, and cultural institutions. A reaction to this determinism came from the possibilists who saw the environment as relatively passive, with man as the active agent. They argued that within the limits set by the environment, people made choices which were relatively independent of it. A more moderate position of environmental probabilism has recently developed. This recognizes the effects of environmental factors on culture and lifestyle, but stresses that such effects are seldom direct or obvious, but are mediated by cognitive, perceptual and pre-existing cultural variables.

There are two ways by which the environment can alter how people live. Change can result from the objective state of the environment, or from the subjective perceptions of environment. For example when British people were first transported to Australia, a completely unfamiliar environment was encountered. Numerous objective factors to be found in the Australian environment, such as harsh climate, a sparse population, different agricultural potentials, etc., led to some extent to aspects of lifestyle which today seem uniquely Australian. But within this overall Australian culture, various subgroups have been able to maintain diverse perceptions of the environment. The coast-hugging city dweller may well perceive the outback (i.e., other than capital cities) as far more harsh and barren than is the case in reality. Similarly, there seems to be a belief held by many rural and small-town Australians, of the inherent sin and corruption of urban areas, and of the undeniable uplifting effect on the soul of man by living near the land. The concepts of the moral benefits of living on the land follow directly from certain perceptions of urban and rural environment.[2]

It is, to a large extent, the varying perceptions of the environment which determine the influence of environment upon the lifestyle of people, and this perception is determined largely by pre-existing cultural factors. For example, to a European arriving in Australia, the landscape may appear barren. To an Aborigine, however, the environment may abound in meaning, with signals of a physical nature, such as where to find food and water, and of a spiritual nature, such as ceremonial places and tracks of Dreamtime beings.

The environment communicates to people, and people act, at least partially, according to information received. The environmental clues are interpreted according to pre-existing cultural concepts, or frames of reference, as a form of mediation between the environment and the lifestyle. This form of adjustment of behaviour to fit into the environment can be seen either on the micro level, of a person behaving differently in a pub than he does in a church, or on the macro level of a British cultural group adopting new behavioural patterns when migrating to Australia.

A current example of how the environment affects behaviour or lifestyle concerns childbearing. Most Australians live in urban areas, where spiralling home costs, the increasing prevalence of flats, and urban congestion are factors influencing them to have fewer children. As the environment becomes more densely populated, a culture will tend to shift in such a way as to induce people to have fewer children. Many cultural factors will be involved in this process. For example, five years ago, only a minority of Australians would be aware of global population trends, and very few would suggest that Australia was overpopulated. But the perception of population and the environment has altered dramatically for many people. There is still social pressure on people to have children, but these pressures are being tempered, and in some circles are even being reversed. An American example of this concerns N.O.N., an acronym for National Organization of None parents. Their motto is "N.O.N. can be fun". Each year they nominate a N.O.N. parent of the year (male or female). In 1974, Australia had its first N.O.N. Father of the Year, the award going to a Brisbane conservationist. Certain sections of the Z.P.G. (Zero Population Growth) movement in Australia also propagandize the N.O.N. ideals. Z.P.G. Queensland branch had, for example, as one of its mottos, "U.N. population year—None is fun for '74".

Thus we have a case where the environment is becoming more

populated (objective fact). The perception is one of overpopulation, or at least threatening overpopulation. This has had immediate effects on certain groups who instigate propaganda which gradually alters the cultural concepts of ideal family size, and thus the lifestyle, at least with regard to having children.

There now appears to be empirical evidence of this environmental effect on lifestyles in the United States of America. Paul Ehrlich[3] has argued that most people in the United States understand that population growth is an enormously serious problem. There is, he states, evidence that the campaign to make people aware of population growth has had an impact on birth rates. Data from the President's Population Commission report indicate a correlation between the number of children that people had and their expressed concern over the population problem. Thus there appears to be evidence that the perception of overpopulation in the environment is beginning to alter the lifestyle, at least with regard to family size. This will then have a further environmental impact, of a less rapid rate of population growth.

Recent experience points up the remarkable cultural adaptability of mankind to environmental changes. For example, much of the population increase of the nineteenth century occurred in conditions which many of us would regard as unbearable. Today, people are achieving a form of physiological and sociocultural adjustment to forms of pollution and stress which are experienced in industrial and urban environments. But such cultural adaption may not be good and may, in fact, portend greater future problems. All over the world, the most polluted and crowded cities are also the ones that have the greatest appeal and where population is increasing most rapidly. Conditions that appear undesirable biologically do not necessarily constitute a handicap for economic growth, in fact the opposite would seem more common. Great wealth is produced by men working under high nervous tension in atmospheres contaminated with chemical fumes, or in crowded offices polluted with tobacco smoke.

René Dubos[4] argues that "biologically speaking, adaptability is almost by definition an asset for survival. However the very fact that man readily achieves some form of biological or social adjustment to many different forms of stress is paradoxically becoming a source of danger for his welfare and his future. The danger comes from the fact that this adjustment commonly results in pathological consequences; but these pathological effects are often delayed and extremely indirect."

In this section, several examples were provided of how environment can determine the lifestyle of people. That there is such a relationship seems consistent with common sense. It appears that man is quite adaptable in the way he lives, and that environmental effects such as pollution and overpopulation may be integrated into, and allowed for in, the behaviour of people. Such cultural adaptability to environmental alterations may in fact reduce the chances for long-term survival of the human species. Lifestyle may in fact be too easily subject to environmental influences!

LIFESTYLE DETERMINATION OF ENVIRONMENT

In discussing the effects of the environment on people, it is often implicitly assumed that they are placed in an environment which then acts upon them. But in reality both animals and people will select and modify their environment whenever they are able to do so.

Throughout history, man, the cultural animal, has altered his environment[5] by the use of fire, adoption of agriculture, and domestication of animals. As various cultures evolved in different ways, each with markedly different lifestyles, so their environmental impacts also varied widely.

That modern man has altered his environment in untold ways is obviously true. A 1970 study[6] estimated, for example, that man is exploiting 40 per cent of the earth's land area, and that he has reduced the mass of land vegetation by one-third. Mankind now consumes about 10 per cent of all the atmospheric oxygen every year in his cars, power plants, and other industrial needs. Similarly, mankind has increased the atmospheric concentration of carbon dioxide by 10 per cent since the turn of the century. The flows of many metals and chemicals through industrial society exceed the natural flows of these materials through the biosphere. The degree of effect which man has had on his environment is truly profound! But it is not man the animal who has so altered the environment, but man the social being. It is man within a given cultural context who takes a certain stance in regard to the environment, who follows a certain lifestyle, and thus who will alter the environment in a particular direction.

One very interesting example from recent history, of cultural concepts and the resultant lifestyle having profound environmental effects, concerns the patterns of roads and land holdings imposed upon an area, the types of structures built, the massing of building in cities, the determination of where people will live, and even how much land a city will cover. For example, there is only a weak relationship between the population of a city and the area it covers. The Australian ideal of owning one's own home, on 0.24 or 0.30 ha, with a fence around, and as much space as possible between one house and the next, is a purely cultural perception of the ideal lifestyle. Such a "Cult of the Cottage" is very wasteful of land, creating an urban sprawl which devours millions of hectares of farm land. The cultural desire for a lifestyle which entails such low-density housing also renders efficient public transport systems unlikely. Thus, there follows the environmental impact of freeways, devouring 9.72 ha of land per kilometre and 3.24 ha per interchange, high energy demands, pollution from car exhausts, and other direct environmental effects. A lifestyle with, perhaps, a greater need for, or less fear of, close proximity would have a totally different environmental impact.

Another way by which cultural concepts can have profound environmental effects concerns family type, size, and population growth. Although there are many diverse cultural determinants of population growth, one series of research projects[7] has found that a woman's acceptance or rejection of the feminine stereotype in her culture largely determines both the number of children she desires and the number she will have. It was found that women who held relatively masculine self-concepts had significantly smaller completed families than did women who held relatively feminine self-concepts, the terms masculine and feminine being defined by the women themselves. This would suggest that the cultural phenomenon of Women's Liberation, to the extent that it aims to alter feminine self-concepts and partially replace them by masculine self-concepts, may well alter lifestyles significantly enough to have profound effects on the population growth rate and, thus, on the environment.

D.H. Stott[8] has pointed out many other ways by which the lifestyle of people affects their rate of population growth. He argues that the most effective devices for limiting human population numbers are cultural. Infanticide would be the most obvious self-conscious control of numbers. In many traditional cultures, the taboo against sexual intercourse during the prolonged nursing period has the effect of spacing out births. The customs and sexual morality of institutionalized marriage also have the result of limiting the number of children which could otherwise have been born. The generally observed custom in many tribes, by which the women are given the heavy work, is a further example of a cultural or lifestyle factor which would reduce births.

There are in fact many ways by which lifestyle will affect population growth rate. Some of these point toward an equilibrating mechanism by which population levels are stabilized well below the biological limit.[9] From this some authors conclude that the projected population increases leading to overcrowding and starvation will not occur. Lifestyle factors will intervene first. The only question is whether these intervening factors will be conscious and moral, or unconscious and amoral.

There are infinitely many, less obvious impacts of lifestyle on the environment. People who are highly literate, for example, will have a greater demand for paper for newsprint and books than will less literate people. Television and radio, it would have to be assumed, will have partially reduced the demand for paper.

The modern phenomenon of greater social mobility may have an environmental impact.[10] One of the effects of mobility has been to exaggerate man's feeling of being separate and aloof from the natural world. Mobile man may lack a sense of place and be unable to perceive changes in his environment, because he is never in any section of it for long enough. This alienation from nature may create a man who not only does not understand the effects he has on the environment but who may not even care! Not only is it clear that modern man, with his high mass consumption and mobile lifestyle, has a marked effect on his environment, but also it appears that his environmental impact is increasing.

In reality, there is very little of the environment which has not been altered to some extent by man. Rapoport argues that: "Since there are now few places left on earth which man has not altered in some way we could say that much of the earth is really designed ... Designed environments obviously include places where man has planted forests or cleared them, diverted rivers or fenced fields in certain patterns. The placement of roads and dams, of pubs and cities are all design ... The work of a tribesman burning off, laying out a camp or a village and building his dwelling is as much an act of design as the architects' or planners' act of dreaming up ideal cities or creating perfect buildings.

What all this activity has in common is that it represents a choice out of all the possible alternatives. The specific nature of the choices made tends to be lawful, to reflect the culture of the people concerned."[11]

CONCLUSION

We have seen that there are diverse relationships between the lifestyle of people and the environment within which they live. These relationships ranged from the obvious to the subtle, though two trends were observed. A culture tends to take signals from the environment and to change lifestyle in any aspects which clash with environmental reality (though this reality is as perceived). On the other hand, any culture creates forces which will try to alter the environment to make it more agreeable to the lifestyle. Over time, if there were no other factors operating, the lifestyle and environmental subsystems would move ever closer to agreement or equilibrium.

But this idyllic state of man living in harmony with nature has rarely, if ever, been achieved. The reasons for this failure to achieve an equilibrium state are twofold. Firstly, both lifestyles and the environment have always been subject to evolutionary changes, as discussed in chapter 11. Secondly, both the environment and the lifestyle are always subject to other influences, such as ethics and technology. Conflict situations are set up, and changes may develop. For example, technological change may well alter the lifestyle in a direction of environmental conflict, quite opposite to the direction expected from a study of environment-lifestyle interactions only.

These various forces have always been present and active throughout the history of man, and conflict or tension between the various subsystems has been the usual state of the overall system. Thus while the subsystems of lifestyle and environment may display tendencies toward agreement or equilibrium, such tendencies may be overcome by the destabilizing influences referred to above. But what is so unique about the present period is that the naturally occurring, equilibrating tendency between the environmental and cultural subsystems may well be overwhelmed by forces of rapid technological development. A certain amount of change can be tolerated, but at some point the change is too great in various subsystems, and the overall system may well break down.

In fact, it has been argued that the whole environment-culture change mechanism is now beyond control. "Our society is trapped, our decision makers do not make decisions."[12] This is somewhat the same point Eric Fromm makes in *The Revolution of Hope*, where he argues that "Perhaps its [present society's] most ominous aspect at present is that we seem to lose control over our own [social] system."[13] Charles Reich in *The Greening of America* produces much the same argument for the inevitable breakdown of the environment-culture mechanism; in a chapter entitled "The [social] machine begins to self-destruct", Reich argues that "It is the very essence of the Corporate State that no one can control it ... the fact is that our society seems incapable of doing the simplest and most obvious things to save itself."[14] Many of these "simplest and most obvious" changes are the lifestyle solutions discussed in chapter 3.

In this chapter, then, we have examined some of the many ways in which environment will affect the lifestyle of people, and in turn how the lifestyle will affect the environment. It was pointed out that in general these two subsystems will tend towards equilibrium, though the environmental crisis may well present factors which override this equilibrating mechanism and lead to disparities of the types suggested in *Future Shock*.

8 Relationship between ethical and environmental subsystems

At first glance the relationship between ethics and environment may not appear obvious. We tend to perceive these as two distinct areas of knowledge and understanding, with little interaction. For example, there is generally little cross-fertilization of philosophy and biology departments within a university. Yet the fact that ethics and the environment are typically seen as nonrelated may owe more to the reductionist philosophy of education than to objective reality.

In a less developed culture, the environmental determination of ethics may appear reasonably obvious, although the ethical determination of the environment may appear less clear. That the ethical position of modern man can have profound environmental effects is due to our scientific and technological powers. In such modern developed societies, we have so dominated our environment that the conditions supporting life on this planet, the land we walk upon, the air we breathe, and the water we drink, are now subject to political management on a major scale. Our relationship to nature is no longer solely determined by the forces of nature but also by the rules of political management. Because of this all-pervasive influence that we are now able to exert on the environment, the attitudes and ideas with which we work become inextricably a part of the environment. The crisis of the environment, seen in this light, is largely a reflection of the values which guide the institutions and procedures by which we influence the environment.

The ethics or values to which we subscribe help determine the environment within which we live, and that same environment partially determines those ethical values. In this chapter, the interdependence of ethics and the environment will be explored by means of the examination of several relevant examples. It is important, however, that the model with which this book deals be kept in mind, so that the discussion is slotted into the overall systems framework.

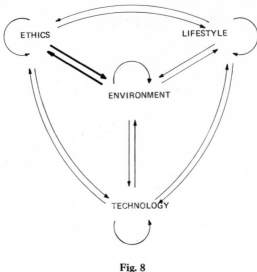

Fig. 8

ETHICAL DETERMINATION OF ENVIRONMENT

Our perceptions of the environment are at least partially determined by the ethical framework within which we function.[1] Historically, such a framework has varied widely, as have environmental perceptions. The idyllic nature portrayed in the writings of Rousseau, Thoreau, or even Wordsworth contrasts sharply with the more pragmatic perceptions of nature held by Francis Bacon, Darwin, or even E. Buckminster Fuller.

But in spite of such variations, the fundamental ethical development of Western man has seen nature as there to serve man. At times nature is seen as an enemy to be conquered, at other times as a benign master, but always there to serve mankind. While the derivation of our commonly held ethical perceptions of nature has been open to dispute, few question that our commonly held concept of the environment is as something quite separate from mankind. Man is perceived as separate from nature, and the correct (read ethical) use of nature is to serve man. This ethical basis provides the framework within which man shapes the environment.

The ways in which variations in this ethically based perception of the environment alter the actual environment provide the best indications of the relationship. The ability to produce a certain type of environment is not everywhere the same. But where wide latitude exists, so too do a wide variety of environments develop. Lawns and gardens around one's home, and even the type of home and city, immediately spring to mind, as examples of ethically shaped environments. The well-shaped British garden, surrounded by shrubs and trees all neatly trimmed, contrasts with the expanse of grass typical in Australia or the United States of America, or with the obligatory solid wall around the French garden, or the small, enclosed, manicured Japanese garden. This same relationship on a broader scale applies to regional and town planning, the type of accommodation we desire, and even the type of political boundaries to which we adhere.

Currently, there tend to be two main types of ethical stance towards the environment. This dichotomy has the nature of being all-inclusive, and the holding of these two belief systems can be found in groups ranging from radical ecologists to industrialists. Briefly the two stances, which are ethical in essence, can be described as: (1) seeing nature as inherently good and worthy of preservation from the ravages of evil mankind (preservationist); and (2) seeing nature as a resource there to serve mankind, and as having value only insofar as it does this (utilitarian or transformational). Obviously, whichever basic ethical stance one works from would determine to a great extent how one treated the environment, at least in the long run.

The utilitarian ethical stance with regard to the environment is epitomized by blunt statements such as "The proper use of water is to dirty it",[2] or, "We must relate any problem of pollution to human purpose, else there is no problem",[3] or more subtly, "We need the [Great Barrier] Reef, all of it, because it is the only one and because we do not understand it. We need it to look at, to touch, to feel, to learn from ... Our need to conserve the Reef is founded in the human condition and conservation of the Reef *for* mankind is the only responsible way we can act."[4]

The preservationist ethical stance in relation to the environment is epitomized by statements such as "Nature, the non human world, in addition to having instrumental [useful to man] value has an intrinsic value in itself".[5] Professor Charles Birch goes on to argue even for legal rights for natural objects such as trees and animals. He totally rejects the man-centred value system which underlies the utilitarian conservation movement. A now famous minority opinion of the United States Supreme Court, from Justice Douglas, suggested this very principle, that natural objects, i.e., the environment, should have legal status.[6] In effect, this requires a radical shift in the way in which the environment is viewed. It suggests the environment is not there to serve man, but that it has rights. Such a statement, if criticized as reification, indicates how ingrained are our ethical concepts of the environment.

It is quite obvious that the long-term treatment of the environment will differ markedly depending on which of these two ethical stances one holds. A rapacious land developer or clear felling forester exploits the environment for his own ends (and may well do so very inefficiently). The regional or town planner or the forestry official will put controls on such exploitation, but the object is to increase the benefit to humans and not save the environment per se. Many conservationists go one step further, perhaps using a spiritual argument, and state that the resource must be preserved for the ultimate well-being of mankind. But all of these environmental impacts are justified by their ultimate value to man.

Dr. J.G. Mosley,[7] director of the Australian Conservation Foundation, has argued that such a utilitarian frame of reference for environmental planning might only result in the more orderly deterioration of that environment.

It is interesting that quite a number of active conservationists seem to adhere to the preservationist ethical stance, yet rarely argue in these terms. "We must be practical" is the well-worn cliché which induces most conservation arguments to be phrased in the language of the utilitarian mind. It is not politic to talk of the inherent beauty of an area, or of the value of a natural object per se, without any particular reference to man. This switch from the actual ideology held to that ideology which seems more acceptable to the majority leads to some interesting results.

An interesting example of this switch in ideology by conservationists applied to the campaign to save Texas Caves which were in danger of inundation from water behind the proposed Pike Creek Dam in southern Queensland. The reason for the conservationists' opposition was almost purely based on preservationist ethical beliefs. Speleologists almost invoked a reverential tone of voice as they talked of the wonders of cave formations and the marvellous intricacies of underground caverns, passage-ways and flooded tunnels. Yet the line of official conservation argument[8] was almost purely utilitarian. An economic benefit-cost study of the proposed irrigation scheme was prepared by several economists. The various effects on agriculture and grazing land along the Dumaresq (Border) river system were evaluated by several agriculturalists. Even the economic value of the tourist potential of the caves was compared to the economic value of the tourist potential of a large artificial lake. All in all, the study prepared by the conservation interests was of a higher quality, more comprehensive, and more in line with generally accepted feasibility principles than was the official study prepared earlier by state public servants. (As I was slightly involved in both studies, I feel qualified to judge.)

Yet, in truth most of the conservationists were not particularly concerned about the economics of the programme, knew and cared little about agriculture in the Dumaresq valley, and many would be loath to encourage tourists into the Texas Caves because of the damage of thousands of tramping feet, of cigarette smoke, and kids' palms. Their motives were largely preservationist, yet the political approach was utilitarian.

The relationship between such an ethical stance and the ensuing environment may seem clearest in the case of a society alien to ourselves. There is a danger that such profound relationships between our ethics and environment may be lost to our view, that we may not see the forest for the trees. But such an oversight does not negate the fact that the environment is to a large extent shaped by the ethical framework within which people perceive the environment and their position within that environment.

In recognizing the intimate relationship between ethics or values and the environment, several authors propose schemes to alter these values through the education system, and thus reduce man's environmental impact. "This flood of new knowledge [about the environment] increases the possibility for increased environmental concern, but it does not guarantee it. Unless the current value systems deem this information important, it will produce little increased concern ... In our society, where some education is compulsory for all, our schools offer one opportunity to engender the attitudes and values which can lead to more responsive environmental problem-solving. The principal feature of the philosophy of environmental education is that man is an integral part of a system from which he cannot be separated. Specifically, this system consists of three components—man, culture, and the biophysical environment."[9]

While this sort of change in ethics might be argued by some as moving people toward a more realistic assessment of reality, others may see in it a form of brainwashing. Some authors are quite blatant in their proposals to alter environmental ethics through education. Professor Meredith Thring, head of mechanical engineering at London University, when on a lecture tour in Australia distributed 10,000 copies of a 130-word pledge, by which students would swear to work only for peace, human dignity, and self-fulfilment. Professor Thring described this as a Hippocratic-style oath for scientists. He urged students to turn themselves into a fifth column in the community to preach the evils of materialism.

In a proposal for a Survival University[10] the author argues that the biology department must point out that it is sinful for anybody to have more than two children, the engineering student would learn that it is immoral to build certain environmentally harmful structures, and the earth science department would teach that it is wrong not to live the simple life. The whole aim is to force a student to feel twinges of conscience.

As well as being morally questionable, it is doubtful if the relationship between ethics and the environment is of a simple enough causal nature to justify such blatant manipulation.[11] The very object of this book is to argue against such simplistic environmental solution proposals by pointing out the elaborate complexity of the interrelations of a host of factors, and not just ethics and the environment.

ENVIRONMENTAL DETERMINATION OF ETHICS

This half of the broad spectrum relationship between environment and ethics can best be conceived of on a two-plane, micro and macro, level. On the micro level could be considered such questions as the type of urban environment and architecture as it relates to certain ethical values of the community, effects on moral values of the man-made, physical environment of institutions such as a prison or mental hospital, and even questions such as the type of physical work environment and the sort of values it may induce or promote. On the macro level is the sort of historical, culture-wide, or cross-cultural comparative analysis of such a relationship. Questions such as the extent to which puritanism was shaped or refined by the vastness of the American frontier, or the effects on British values of migration to the vast, hostile Australian environment are here relevant.

On the macro level the clearest, simplest example of the environment shaping ethics is provided, perhaps, by mass migration movements. With regard to the early mass migration movements from Europe to Australia, the people who came brought with them a set of concepts of the environment, and these concepts were intimately related to the purpose or goal of their migration.[12] While some saw the environment as a source of wealth to be exploited as quickly as possible, before returning to civilization, others saw the environment as a gaol without walls. The mass migration movements transferred not only people but also a whole set of spiritual and ethical values from one side of the world to the other.

Heathcote categorizes the various spiritual perceptions of the Australian environment as scientific, romantic, colonial, national, and ecological. While the scientific, romantic, and colonial perceptions of the environment tended to be imported, deriving from the

English environment, the national and ecological perceptions developed in Australia. For the migrants did not find an environment such as that in which their value systems had developed.

When incongruity exists between the value system one holds and the environment, there are really only two options, though these are seldom conscious. Either the environment or the values will be changed. Pressure will be applied in order to alter the environment, to mould it more similarly to the environment in which the value system developed, or the value system will change to be more in agreement with the new environmental realities. The latter is the more common result, though both trends were obvious with the first migrants to Australia, and are still in evidence for more recent migrants.

A desire to change the environment to fit into another value system was seen for example with the introduction of European flowers and trees to Australian gardens. Experimental Farm Cottage (circa 1800) in Parramatta had a very typical, small, enclosed English garden in what was then the bush, with almost every plant species introduced from England. Similarly, housing styles often paid more attention to imported ideologies and styles than to environmental realities.

An overriding concern in early Australian town planning was the desire to look like home. Even today one often hears Australians remark with pride how much Melbourne resembles an English city, or Victoria resembles the English landscape. But, as architect and historian J.M. Freeland notes, "Despite the conscious attempts to recreate a little bit of Home—the replacing of indigenous flora with English exotics and the attempt to build English style architecture housing, English furniture—the imperatives of climate, topography, and economy forced them to come out Australian".[13] Attempts to alter the environment to force it to agree with a preconceived value system more often than not have resulted in changes in that very value system. This is the most common response to a conflict between values and the environment, i.e., to alter these values.

Of course the *Bulletin*, and the pens of Henry Lawson, Banjo Paterson, Norman Lindsay, and Bernard O'Dowd were equally based on spiritual or ethical growth within the Australian environment. This type of nationalist fervour was also shown in the political ideologies and in directions taken by politics. Here too the environmental effects were obvious. "Politicians seem to have had varying attitudes to the continental environment, but all, from the optimists to the pessimists, from the far right to far left, saw the resources of the land as the basis on which the particular society they had in mind was to be founded."[14]

It appears now, however, that changes in the environment, or more often changes in the perception of the environment, are leading to a new set of ethical values. This is the area Heathcote calls "the Ecological Vision." Such an ecological vision has developed on a world-wide basis over the past few years. While concern for the environment is not new, there are several unique features in the present conservation movement which mark a new era, the main difference being the change in basic values or ethical stance of many adherents. The early high prophets of the new (radical ecologist) conservation movement would be people such as Rachael Carson, Paul Sears, Barry Commoner, and Paul Ehrlich. They described in layman's terms the intricacies of ecology, the web of life. They warned us of dangers in the way we pollute, exploit our resources, and breed with abandon. Though the warnings were generally phrased in the utilitarian manner previously discussed, the ethical stance engendered was more preservationist.

Out of the environmental movement has grown a change in ethics which many early proponents likely did not foresee. The ecological vision is one of man not as a dominant species, but as a link in a biological masterpiece. Traditional concepts like man's mastery of nature, economic growth and the idea of progress, and even firmly held ideas of good and bad all come to be re-examined in light of a new radical ecological awareness. This becomes then a new philosophy of life which separates the new radical ecology from the old style conservation movement. Radical ecology is a system of beliefs resulting from a certain perception of the environment and of environmental change. It sees the environment and man's position therein as much more than a biological fact, as having spiritual significance.

One of the most interesting social developments in Australia today is the large Tuntable Falls Commune near Nimbin. This is a 423 ha property on which eventually several hundred people may live, in several villages, providing a rural home for the counter-culture movement. The religious crusade aspect of their ethical stance is apparent in much of the early literature which poured forth. "To live in harmony with nature and with ourselves with love and understanding. To be free from pollution of air, food, bodies and mind. To create a society on these grounds for the betterment of other societies and the world ... We ask only that they [visitors] leave, feeling that they have helped or been helped in the struggle for survival and the creation of a more beautiful world."[15]

This is an ethical stance which has developed from an environment perceived in a quasi-religious light, an environment they see as desecrated by an evil social system from which these people wish to dissociate themselves. The newly popular ecological morality then is very much an example of environment altering ethics. Such an ethical stance is not new, though the popularity has only recently been on the increase.

CONCLUSION

Several examples have been used to show how ethics help shape the environment and the environment helps shape ethics. Given a lack of other influences, these two subsystems within the heuristic system being developed will tend to move toward equilibrium or agreement. But, of course, there are many other factors working at the same time. While this chapter has discussed several of the myriad relationships between ethical beliefs and the environment, showing each subsystem as influencing the other, it is important to bear in mind that the relationships are generally not of a direct causal nature.

In summarizing empirical research relating environmental attitudes and behaviour towards the environment, Bruvold concludes: "Predictions involving relationships between environmental attitudes and related specific beliefs, and between environmental attitudes and related specific behaviors, received [only] moderate support ... These results are consonant with general research in this area, which indicates that, if relations between general attitude and specific behaviors or beliefs are found, they are usually unimpressive."[16]

This lack of clarity is, of course, in agreement with the theme of this book. Such straightforward relationships are confounded by the many other factors. Along this line psychologists have written: "One does not come to a new environment as an inert passive entity. One goes to a new environment with other purposes in mind ... and with a certain projected schema of what the new environment will be like. This projected schema typically derives from certain norms established earlier in life, norms built up in environments in which one has long sojourned."[17] This is the intricate system of influences with which this book deals.

9 Relationship between ethical and technological subsystems

While our ethical values are deeply ingrained and often shielded from conscious thought or public scrutiny, technology is the very opposite, being glaringly apparent to all. The tools we use, from the simplest garden implement to the most complex computer, are part of this all-pervasive repertoire of technology which we have devised and provided for ourselves.

A superficial view would see this technology as simply a set of tools by which man tries to achieve ends which are determined by a process quite separate from the technological tools themselves. While some people perceive technology within a nearly value-free framework, it is more common for technology to be referred to in heavily value-laden terms. Jacques Ellul,[1] one of the best-known analysts of the technological society, believes that technology is no longer simply a means by which we achieve our ends in modern society, but rather that technology has become an end in itself, an end which dominates all other ends, including moral or ethical ends.

With regard to the role of technology in environmental problems and solutions, the tendency to see that technology in heavily value-laden terms is most pronounced. Many authors (see chapter 3) see technology as that which can save mankind from all its problems. In glowing terms, they talk about a new era which will dawn for mankind, ushered in by technology. Environmental problems will be solved, man will work less, be happier, and all will be well. There are many political commentators and journalists who see technology in a very positive ethical light. Benefits ranging from the realization of democratic ideals to greater personal freedoms are all attributed to technological development.

There are many other authors who see technology in heavily value-laden terms, but perceive it as bad. Even scientists as well known as Barry Commoner see in our technology the fundamental cause of our serious environmental problems. Toffler[2] sees in technology a source of personal and social unrest. Heilbroner[3] argues that pressures which are basically technological are causing history to "close in" on the United States, bringing no sudden crisis, but instead a steady worsening of present inadequacies and impoverished philosophies. Along basically similar lines, Marcuse[4] sees technology as the root cause of the demise or end of dissent and major social change. He argues that the technological order is a comfortable, democratic unfreedom, a totalitarian system in which technology replaces terror. Thus, he argues, technology leads to "one dimensional man". Both Reich[5] and Roszak[6] in their analysis of the counterculture movement place great emphasis on the negative ethical connotations of technology. Douglas, in studying a number of such anti-technology ethical stances, refers to "the inherent secret tyranny in the technological society".[7]

In summarizing these two strands of polemic thought in regard to technology, Douglas concludes, "whether one foresees gloom and doom or bright utopias arising from the steady advance of the scientific-technological world view seems at this time to depend primarily upon one's general moral orientation or even one's ideological mood toward our modern society".[8] This logical statement seems self-explanatory in light of the examples which follow.

In this chapter, the relationships between ethics and technology will be explored. Sometimes direct change-causing relationships can be found, but more often, the relationships are diffuse and indirect. This is in line with the holistic concepts of this thesis. Ethics and technology are related in many ways, but other factors are involved as well. It is essential that the heuristic model be borne in mind, and the following discussion be seen in that context.

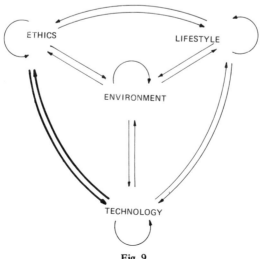

Fig. 9

ETHICAL DETERMINATION OF TECHNOLOGY

To any observer of recent history, one of the most phenomenal aspects must be the all-pervasive changes in our daily lives brought about by technology. Particularly since World War II, this technological change has been a prime determinant of all social change. But why, at this precise period of time? Why in certain cultures while not in others? Why certain technological developments, and not others? Why can we send men into space, yet are unable to cure the common cold, or preserve a clean environment? Why is it that in the past decade in the United States of America less than 1 per cent of all government research investment has gone toward such critical social problems as housing, crime, the urban environment, and ground transportation? Why is it that our technical priorities remain relatively stable, despite the fact that the world about us is changing dramatically?

In trying to understand such questions it is important to understand technology in an historical light, to see it as part of our complex social structure. Technology is born of a certain idea of nature and material as well as nonmaterial progress, and is related to specific social ideals and specific concepts of the proper ends of human life.

Historically man has always had at his disposal a considerable repertoire of technical instruments. As far back as the Paleolithic or Old Stone Age (12,000—20,000 years ago), the axe, the flint scraper, the spear, bow and arrow, gimlet, oil-lamp, and numerous types of bone instruments were in common use.[9] As agriculture developed, these crude stone instruments were refined and improved in various ways, being replaced by bronze, copper, and iron. Also, technology was refined and adapted to ends varying from home comforts such as a flush toilet at Knossos in 1400 B.C., to artillery, to commerce, and even to temple doors which opened automatically when the sacrificial fire was lit.

Over the same period of technical development, the prevailing attitudes toward technology have varied markedly. To the classical Greek mind, this was regarded as a low prestige area of endeavour. To deal with everyday events, to produce and to trade were activities fit only for slaves and foreigners. It was the realm of distant and changeless ideas, rather than the objects of our mundane world, which were seen as worthy of the Platonic mind.

The Roman value system, however, placed a higher merit on technology. To achieve in a physical sense, to dominate both man and nature, was seen as a goal worthy of man. Technology appropriately was developed along military and industrial lines, providing an efficient army, and even facilitating mining operations to a depth of more than 210 m, with provision made for pumping water out, moving ore and debris and even using water under pressure to remove the overburden.

Differences in attitudes toward technology have influenced the development of technology in any particular culture. One of the most important influences is the way in which nature has been conceptualized. In tracing the development of modern technology many authors seek direct explanations in terms of philosophical or ethical bases. Thus, for example, Darling has argued that modern technology has evolved from our Judeo-Christian philosophical background. This was an ethical viewpoint which saw man as on earth to conquer nature, which valued empiricism, and which had a concept of progress which is explained as part of Judeo-Christian teleology. "Our Greek derivation in Western civilization gave us the reason which has guided our science, but the Judeo-Christian background has given us a man-centred world. Our technology is a monument to the belief that Jehovah created us in his image ... The resources of this planet were for man, without a doubt. They could have no higher end than to serve man at the behest of Jehovah. There could be no doubt of the rightness of technology."[10]

Regardless of where we believe the ethical foundations of modern technology originated, it is reasonably easy to see the sort of ethical values people attach to technology today. Everyday newspaper accounts often unwittingly disclose deeply held ethical values. An excellent example concerns an editorial in the *Australian* entitled "It's engineers we need, not more BA's".[11] Engineering is taken as at the crux of technological development. It has been found that while university students are flooding arts courses, many positions in engineering schools remain unfilled. Such a widespread rejection of a career related to technology would have complex causes, but at least partially it would be explained by attitudes towards technology. There are many personal rewards attached to engineering (like plenty of job opportunities and relatively good pay) yet people shun it.

The observation that young people en masse are not fitting into the demands of the technocratic society, that they may see themselves as more than tools for technology, leads the *Australian* to ask "how people have come to get their priorities so desperately wrong?" In the 1969 Reith Lecture, we were warned that technology is apt to condition us psychologically so that man becomes its servant, no longer its creator and master, and this seems to be the very conclusion accepted by the editorial writer. A follow-up letter to the editor, with a generous helping of rhetorical nonsense, was provided by the Vice-Chancellor of Monash University, "although life on earth may be improved by the study of social sciences it will only continue to be possible by devising and applying the best possible technology".[12]

It seems glaringly apparent from such value-laden comments that values or ethical beliefs are closely bound up with technology. The direction which technological developments follow at any one time is to a large extent determined by the priorities and values of those segments of society which wield the greatest economic and political power. That a space race developed between the United States of America and Russia during the 1950s and 1960s is an outcome of the "cold war" with all its attendant beliefs of competition, superiority, competence, etc. That this was a race to the moon, rather than a race to cure cancer or to solve serious environmental problems like over-population or even a race to define and build a truly just society, can be seen as a consequence of the value-laden priority systems of those with power.

Another way to think about this relationship between ethics and technology is to speculate how different our world would be if different ethical values had guided our scientific and technological development. If our idea that more is better was replaced by a striving for perfection, would we have any large factories? If our work ethic was replaced by a leisure-time ethic could we see present essential production maintained on a ten-hour work week? If the idea of technology as the art of directing the great sources of power in nature was replaced by an idea of technology allowing man to minimize his environmental impact, what sort of new energy sources would we utilize? The examples can be listed in a never-ending fashion. What is clear is that we use technology as a tool to accomplish certain ethically determined ends. But because of the value-laden nature of the ends, and the entwined nature of means and ends in real life, the technology becomes to a fair extent a derivative of the ethics. "The values existing at any given time will determine the technological choices that a society will make."[13]

TECHNOLOGICAL DETERMINATION OF ETHICS

In the first part of this chapter we have seen a few of the many ways in which ethics or values not only shape technology but actually become embodied in the technology such that the two are intimately connected. Much the same process applies in reverse, values or ethics being partially determined by technology. Again it is important to remember that simple cause-effect relationships are not suggested, that what is proposed is a dynamic system, an understanding of each part of which is dependent on an understanding of the whole.

The various influences of technology on ethics can be seen on both a micro and a macro level. The micro level refers to certain elements or specific aspects of technology, while the macro level refers to technological change in general.

A very clear example at the micro level, of technology helping shape ethical values, concerns birth control. Various methods of birth control have existed throughout human history,[14] with abstinence, abortion, and infanticide being the most common. Modern technology entered the field with the invention of the condom in about 1560.[15] From that time onward pregnancy has been ever less an inevitable result of intercourse. Still, however, the official Christian morals have assumed such an inevitability. During this century various technological developments have made the official taboo more and more difficult to sustain. The release of the popular birth control pill in 1960, and the more recent development of vacuum, or auto, abortion, have made such an ethical value seem devoid of logic. How can a person argue the case for chastity based on fear of pregnancy, when technology has so changed our biology?

There is always the rearguard action of moral watchdogs. Typically, attempts are made to deny the new technology to all or at least to certain people. Thus, while in many countries condoms are dispensed from small vending machines in public places along with cigarettes and sweets, such availability is denied here in Australia. Some doctors are still loath to prescribe the pill for single girls. Abortion was originally outlawed in England (and consequently in Australia) because of the dangers inherent in the medical technology of the time. Advances in medical technology now make it several times safer to have an abortion than to have a baby, so the original justification for anti-abortion legislation is gone, yet moral guardians fight on with new rhetoric. Popularization of the newest technology of abortion, either small, simple-to-use vacuum tubes which a woman can use herself or with the help of a friend, or prostaglandins crystals, makes the past moral conflicts almost redundant.

What is the chance of an anti-abortion campaign when the abortion takes place not in a hospital, a clinic, or even a neighbour's house but in a person's bathroom, by simply taking a twenty-five cent pill? In such a case, anti-abortion laws become as ignored and unenforceable as most other sexual-legal anachronisms.

What happens is not so much that the old or accepted ethics are changed by conscious effort, but that they are simply ignored, being regarded as quite irrelevant. The well-established practice of parents trying to protect their daughters from sexual involvement, based on fears of pregnancy, simply becomes irrelevant, at least to the daughter if not always to the parent.

Another technological development which is possible, and even probable, and which would have profound impact on ethics concerns sex control. Using various techniques derived from research into fertility and birth control, medical science is moving towards the ability to determine the sex of a child by chemical means. Such a scientific-technological breakthrough would allow a couple to implement currently held preferences for children of each sex.

At a first, superficial glance this appears to be a quite desirable widening of the area of personal choice in procreation. Thus just as birth control technology provides people with the freedom to procreate or not, sex control provides the freedom to have a male or female child if you decide to have a child. Little thought is required, however, to see that such a technological development raises profound ethical questions.[16] While sex control would allow a couple to choose either a boy or a girl, empirical evidence suggests that boys would be the predominant choice. Etzioni[17] quotes several studies, showing 18 per cent to 65 per cent greater demand for males than for females. But, he argues, the actual preference for males is likely to become much greater than indicated by such empirical studies as soon as such a choice becomes a reality.

What will a surplus of males mean? Some delay in the age of marriage of the male, some rise in prostitution and in homosexuality, and some increase in the number of males who will never marry are likely to result. Also, a preponderance of males would increase violence, crimes, interracial and interclass tensions, and some of the rougher features of a frontier town. This form of argument is carried on by Postgate where he suggests among other things, that a shortage of females would lead to a new technology of sex, "Substitutes for normal sex, mechanical and pictorial, would be widely used."[18]

That such a scientific-technological breakthrough would have profound influences on a culture must be quite clear. The fundamental ethical question is whether such changes should be allowed. Given our deep respect for research and our worship of scientific progress, it may not be possible for mankind to stop such research even on ethical grounds. But it is clear that if the technology of sex determination is perfected, as appears quite feasible, such a technology would have truly profound influences on the prevailing value system or ethics.

For much of history there have been those who saw in technology not a saviour but a dehumanizing influence. It has already been pointed out how several Greek scholars regarded mechanical (technological) applications of science as inherently degrading to the nature of man.

In the last century people who expressed this sort of fear of technology per se came to be known as Luddites. Machines were seen as replacing people and violently threatening the social order as well as the physical well-being of workers. Between 1811 and 1816 groups of Luddites viciously attacked new technology, smashing machines, seeing them as the epitome of evil. The term Luddite is still used derisively to denote a person with a deep fear of technological change.

Such negative attitudes toward technology in general have been gradually replaced by the view that technology and science are the new gods. This sort of "religious" respect leads some writers to see in technology the salvation of mankind. Such an ethical stance has wide currency, in the field of environmental problems and solutions. Several examples of such utopian technological values were quoted in chapter 3, as well as being referred to briefly early in this chapter.

But of more interest is a new movement, of what could almost be called the new Luddites. Such people as Herbert Marcuse and Jacques Ellul argue that technology in general changes our ethics or value system in such a way as to harm mankind. This is the macro view, not of any particular facet of technology in isolation, but of technology in general. It forms the essence of a radical critique of technological society and the directions in which it moves.

Marcuse[19] argues that out of the technological society emerges a pattern of "one-dimensional thought" and behaviour in which ideas, aspirations, and objectives are redefined by the rationality of the technological system. And technology by its nature has produced a system which can absorb even the degree of radical dissent which our dehumanized society engenders.

Ellul carries this sort of argument even further, contending that "technology is in process of taking over the traditional values of every society without exception, subverting and suppressing these values to produce at least a monolithic world culture in which all technological differences and variety is mere appearance."[20]

Ellul's explanation for this technological domination of human culture follows Engel's Law, that quantitative change merges into qualitative change. That is, mankind has always been faced with technological change which threatened his value system. But only now the rate of introduction of such changes has increased to a point where, by sheer numerical proliferation and velocity, they surpass man's relatively unchanging capacities to exploit them as means to human ends.

CONCLUSION

In this chapter we have seen various ways in which our values or ethical framework help shape and direct technology. Also, we have seen how specific technological changes lead to ethical changes, as well as how technology per se has tended to take over, in a sense, our whole culture and to shape values.

As with the interrelationships of all the subsystems of this heuristic model, given a lack of other influences, the technological system and the ethical system would tend toward agreement or equilibrium. But in the real world with which this book deals there are other influences which may overpower any equilibrating influence. There are strong arguments to suggest that technological growth develops a momentum, a logic, and even a justification of itself. That, in effect, while still under the influence of other subsystems, there exists a degree of self-sustaining momentum. This is explained in more detail in chapter 11.

10 Relationship between technological and environmental subsystems

The relationship between technology and the environment is one of the least well recognized relationships between two of the elements with which we are concerned. It can be argued that one of the serious underlying causes of the environmental problems we face today is exactly that technology too often has been developed in ignorance of environmental consequences and that, hence, serious environmental damage has resulted from the application of such technology.

As with the other relationships between the various subsystems, simple cause-effect factors are seldom suggested. The relationships and the discussion must be seen within the framework of the overall model in which the various subsystems are interrelated in a complex fashion.

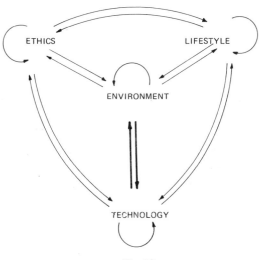

Fig. 10

TECHNOLOGICAL DETERMINATION OF ENVIRONMENT

Mankind's altering of the physical environment by technology occurs in one of three ways: deliberate, associational, or accidental. Deliberate influences refer to examples where the primary human goal of the technology is alteration of the environment. Many examples of agricultural technology, such as the drainage of swamps and the irrigation of arid regions, are deliberate environmental changes.

Associational environmental influences are predictable, though not the primary intent of those utilizing the technology. Thus the building of an urban freeway will have many associational environmental effects such as relocation of "desirable" and "undesirable" urban areas, changes in run-off of rain water, and even shifts in where the city will develop and stagnate. Similarly, introduction of Western medical technology to a new area will lead to associational environmental effects such as increased urbanization, population increases, shifts in age and sex ratios, and other demographic factors, even though these changes may not have been the intent of those who introduced the technology.

Accidental environmental influences are unintended and generally unknown to the people concerned at the time of introduction of the technology. These accidental environmental effects may be seen as either beneficial or harmful to mankind. A beneficial example concerns medical technology in China which strives not to cure but to prevent disease by alleviating water and air pollution. A cleaner, more healthy ecosystem is not the goal but is the accidental result.[1] A harmful accidental example is provided by the agricultural technology of irrigation. In some cases land has been ruined by salination resulting from irrigation. In other cases, the ecology has been so altered by irrigation that insects, birds, and diseases are promoted from a coexisting to a competing status.

The deliberate and the associational environmental effects are generally less serious than the accidental ones, with regard to long-term environmental effects. This is because the former are at least relatively predictable and are generally reversible simply by stopping the offending technology. The accidental effects, however, are normally much more serious. Because they are not expected there is generally no way that people can prepare measures to alleviate the harmful environmental effects. Also, because they are accidental, the environmental effects can often not be reversed simply by removing the offending technology. Examples of the latter point range from diffusion of DDT through all biotic forms of the world, to deterioration of the ozone layer of the atmosphere by the use of aerosol sprays, to eutrophication of streams by a build up of organic matter.

Because of man's increasing knowledge of ecology there are ever fewer legitimate cases of accidental environmental effects resulting from technological changes. Most such changes come to be knowable in advance, at least to some extent. But though environmental effects are to an increasing extent predictable, and thus not accidental, the predictions are too often ignored by those with the power to make decisions and, thus, to them the environmental effects may well appear accidental. The purpose of environmental impact statements is to change environmental effects from accidental to knowable and predictable. Then a rational decision to proceed with or cancel the project can be made, based on sound predictions of the environmental changes entailed. Unfortunately, in practice, environmental impact statements have often been open to abuse and neglect.[2] For example, an American study has shown that most legally required environmental impact statements, even when warning of severe dangers, did not alter the original development plans at all.[3]

Historically, man has long had an influence on his environment, because of the technology he used to feed and clothe himself. Even as a hunter-gatherer, there is evidence that man's technology had environmental effects still obvious today.[4] For example, the use of fire to drive prey by ancient man has shaped the structure of vegetation in areas long used by hunters and has served as a selective force in the evolution of woody plants and grasses.[5] There have been suggestions that such technology is at least partially responsible for the treeless plains in North America.[6] Fire would be the first advance in technology which altered the natural environment wherever it was used by man.[7]

With the development of agriculture and sedentary populations, the effects of man's technology increased markedly. Animals were domesticated and in the process were genetically altered. Genotypic factors, those which are inheritable, were selected for, since only the animals displaying such traits were allowed to breed in captivity. This selection process of early man applied to many domestic animals. Similarly, many plants were domesticated and altered genetically by even the earliest agricultural technology of man.[8]

The early development of agriculture often utilized irrigation as a form of technology. Throughout history there have been the joint irrigation problems of salination and sedimentation. These are thought to have so altered the environment of ancient Mesopotamia as to have contributed to the eventual destruction of the social system which evolved there.[9] These same factors are blamed for the development of desert areas such as the Sahara and the Rajputana.[10]

Modern agricultural technology with its vast increases in power available to it has far greater environmental effects. The provision of irrigation channels, drainage of swamps, aerial seeding of desired plant species, as well as of pesticides and fertilizers, are all examples of gross environmental changes resulting from modern agricultural technology.

More subtle environmental effects result from the use of computers to store and analyze data in selective breeding programmes for animals and plants, and the use of certain types of fertilizers which over time alter the soil composition. While such changes may at first appear minuscule, over longer periods the plant or animal species or even the soil may be radically changed.

In Australia today the most extensive man-made environmental changes result from agricultural technology. This technology has resulted in introduced species of plants such as clovers and grains, introduced animals such as cattle and sheep, and introduced chemicals such as DDT and inorganic nitrogen fertilizer. The massive changes in land forms for dams and irrigation works have greatly altered the geography of areas such as the Ord River, Murray and Murrumbidgee Valleys. All these changes have altered the Australian environment so as to make it significantly a product of human effort. These environmental changes in turn have other effects throughout the environment. The use of nitrogenous fertilizers, for example, will promote plant growth in run-off streams, hinder growth of aquatic animal life, and may even present a health threat to humans.[11]

Some of the most marked technologically induced alterations of environment have occurred where Western technology has been introduced into the environment of less well-developed areas.[12] Very often a form of technology is developed within a certain environment, and consequently the environmental effects are within reasonable limits. Serious problems then arise when the technology is transposed to a different environment.

Technology has been argued to be the major cause of the environmental problems which we face today. Barry Commoner[13] in particular has put forward this argument. In analyzing post-World War II production figures, he reveals empirically that increases in overall production have been relatively small compared to shifts in the types of things produced and in the methods of production, i.e., the technology. The introduced technology, in most cases, has had far greater environmental impact than the technology which it replaces. Thus detergents replace soap, synthetic fabrics replace wool and cotton, nonreturnable replace returnable containers, and aluminium replaces steel and wood. In each case, Commoner argues, there is an aggravation of the environmental damage. Thus, it is not the actual growth of population or the growth of the economy which leads to the major environmental changes experienced, but the direction in which the technology has developed. Commoner also argues that agricultural technology, dependent on inorganic chemicals and intensive land use, has been one of the prime factors in the deterioration of the rural environment, as well as the source of much serious water pollution.

Arguments along somewhat similar lines, linking more specific changes in technology to changes in environment, have been made in numerous areas. For example, cars and highways as a technology of transport,[14] and flush toilets and sewers as a means of handling human excreta[15], have been seen in this light.

While in the past man's technology has had profound environmental effects, and while this change process has increased markedly with the developments in contemporary technology, the future may be far more startling. Futurologists such as Hermann Kahn[16] and Buckminster Fuller[17] predict technologically produced environmental changes of a nature quite unknown to us today. These predictions range from extensive farming and mineral exploitation of the oceans, to placing very large, reflective mirrors in space to increase sunlight and day length on earth, to the development of genetic architecture, i.e., the ability to predetermine the genetic composition of plants, animals, and even man. One of the most feasible future technological developments which would obviously have profound environmental effects would be the control of weather patterns.

Planetary engineering is the collective term used to refer to proposals for massive environmental alterations, brought about by modern engineering technology. The huge Aswan Dam already markedly alters the environment of north east Africa, as well as the eastern Mediterranean. Other proposals concern rerouting water now flowing to the Arctic Ocean so that it irrigates vast areas of southern Russia. Several new lakes, one as large as Great Britain, would be formed. The scheme is expected to alter the climate of western Siberia, allowing agriculture to move north. Cooling air masses above the lake might alter the climate over the fertile steppes to the south of the project.

An even more grandiose technological scheme involves proposals to dam the Amazon River, creating an inland lake or sea one-third the size of France. Flooding of the Sahara Desert and the inland of Australia as well as damming the Bering Strait are all seriously being considered. Nuclear power has been suggested for provision of new canals, including a sea level Panama Canal, allowing water and biotic forms to mix from Atlantic to Pacific. The environmental changes wrought by any such technological scheme would alter the environment of continents, if not the whole world.[18]

ENVIRONMENTAL DETERMINATION OF TECHNOLOGY

Technology is shaped or directed by environmental factors mainly by limiting which technology will or will not work. Environmental barriers or constraints on technology generally take the form of negative influences. Such factors as resource availability, energy sources, and the presence or absence of cold, heat, high winds, or water all help determine the type of technology which is developed and which is adopted and passed on. Though of less importance, technology is also shaped by the environment, when the aim of the technology is to overcome some environmental restraint on another activity. Thus, for example, in Holland technology has been developed over many centuries to drain the marsh areas which were inundated by the sea, cleanse the land of salt, and make it agriculturally productive.

Historically, it is quite easy to see ways in which technology has been shaped by the environment. That steam power, based on the burning of readily available coal, developed in England, while water power was more utilized in eastern Canada, where water is plentiful and rapid-flowing, and windmills came to dot the early Australian scene, can all be seen as forms of technology developed in response to the environment.

Perhaps in agriculture it is easiest to see where technology is shaped by environment. Throughout the world different farming methods are utilized. While economic and cultural factors are important, so too are the environmental factors.

That irrigation technology developed in areas with low rainfall yet adequate surface water, e.g., Mesopotamia and the Nile Valley, should not be a great surprise. That quite different agricultural technologies were adopted during the same period by early pioneers of Australia and Canada can be seen to be largely due to environmental factors. Within Australia quite different farming methods developed along coastal fringes and in the drier, more harsh interior. Soil types and conditions also helped shape the agricultural technology. Where such environmental factors have been taken into account, the agricultural technology will be reasonably successful. In other cases, where environmental factors are ignored and incongruous technology applied, e.g., sugar cane growing in the Isis area of Queensland,[19] problems, both physical and social, are many and of a serious, long-lasting nature.

Many environmental scientists argue that one of the major reasons for current environmental problems is that technology has too often been developed without reference to environmental constraints. The realization of this has recently led to a whole host of new technological developments to help fit into these environmental constraints. A classic example of this is the development of electrostatic precipitators in chimneys to reduce air pollution and the fallout of ash.

Development of technology to fit into environmental constraints is, in fact, one of the fastest growing areas of industrial research, and the application of such technology represents a rapidly growing economic activity in Australia today. For example, an exhibition in Sydney in February 1973 had more than two hundred exhibitors, all of whom were in some way wishing to supply or utilize technology to help fit into the environment. Many of the companies which displayed new technologies to control environmental damage, however, are also some of the worst polluters.

These technological developments are generally of two types. The first type is physical, either modifying environmentally harmful technology, or providing new machinery to reduce the pollution in some way. Also, new technologies are being developed to enable the utilization of resources which are now being made scarce through overuse. Examples of this range from technology to drill oil wells far out to sea to suggestions for mining minerals on the moon.

The second type of new technology involves measurement of environmental effects and predictions of these, so that the offending technology can be modified. The very involved process of preparing environmental impact statements is just one example. Such technology is in its own right becoming big business. There have been numerous charges,[20] however, that all of this technological development, ostensibly to alleviate environmental damage, is really just another form of industrial growth, that industry is combating waste and pollution so it will be allowed to waste and pollute more.

Whatever the fundamental motivation for such environmental technology it seems likely that it will increase in the future. There are many examples of technologies which have ignored certain environmental constraints, and which may well come to be replaced in the future. For example, there have been numerous proposals for, and predictions of, farming the sea, or aquaculture, to help feed the world's growing population and relieve environmental pressures on farm land. New sources of power, ranging from solar to wind, tidal, and geothermal, have all been proposed, to counter the depletion of fossil fuels. Technological proposals have even been put forward to limit the breeding capacity of mankind, in a desperate attempt to control human populations and the resultant environmental pressures.

CONCLUSION

In this chapter we have seen how technology (the environment are intimately related. Examples were provided to show how various technologies in the past have altered the environment, leaving results still observable today. These technological changes to the environment were either deliberate, associational, or accidental, in the intent of those applying the technology. Agriculture, in particular, was shown to have always had a marked impact on the environment. The environmental effects of modern industrial technology were discussed and it was pointed out how technological alterations of the environment in an accidental way should become less frequent as the work of environmental impact statements becomes more sophisticated as well as more commonly used. Unfortunately though, changing a harmful environmental effect from the realm of accidental to deliberate does not at all ensure that the offending technology will be stopped. Thus to understand the intricate links between technology and the environment is not sufficient to prevent environmental degradation.

For the future, the potential threat of such massive environmental changes as those suggested by the technological ideas of futurologists is truly startling. The environment may well be altered beyond imagination. Because of our imperfect knowledge of the workings of the global ecosystem, there is a real danger that our environment could be rendered less hospitable or even inhospitable to human life. Threats to the ozone layer from supersonic flights, and the danger of either a greenhouse or ice age effect on climate are two of the most ominous large-scale, technologically induced problems.

Examples were also discussed showing how the environment acts as a constraint on technology, thus providing direction for its development and application. In particular, the current environmental problems are inducing many new technological developments. This is a trend which can only increase in the future as resource depletion becomes more of a problem, population pressures continue to mount, and people become less tolerant of pollution on either health or aesthetic grounds.

Over time, and in the absence of other influences, it would seem logical to expect the environmental and technological subsystems to move toward agreement. As mentioned earlier, many environmentalists argue that the lack of congruity between environment and technology is the fundamental cause of many environmental problems.

But the self-induced and self-sustaining development of technology (chapter 11) in particular may overcome the tendency towards equilibrium. It may be argued that many elements of current technological development in fact move in environmentally destructive directions.

11 Self-induced changes within subsystems

Chapters 5–10 have discussed the relationships between the various subsystems of the model, and causal or influential relationships were pointed out by means of exemplification. In each chapter, the discussion of the cause-effect nature of the relationships was qualified with the warning that the entire pattern of relationships must always be borne in mind. In other words, all the subsystems interact, though at any one point only two were the focus of the discussion. The subsystems were described as being in a state of continual change due to their interaction with each other subsystem.

In this chapter, I wish to discuss the ways in which the subsystems may change in a self-induced or self-sustaining manner. There are, within each subsystem, elements which tend towards stability or equilibrium, but there are also elements which tend towards instability or disequilibrium, i.e., change. This would seem consistent with systems theory, as well as commonsense observations of the subsystems discussed herein.

As with the other chapters describing changes within the subsystems, it is important to see this discussion in the framework of the heuristic model and not as separate, discrete relationships or change factors.

SELF-INDUCED CHANGES IN ENVIRONMENT

All aspects of the natural environment are in a state of change over time—of becoming rather than being. Given the relatively short time perspective of each of us within our environment, however, we tend to see this environment as a constant. Even the sort of environmental changes with which we must accede, such as the changing seasons, we poetically interpret as a constant, and to this constant rhythm of nature man ascribes mystical and romantic significance.

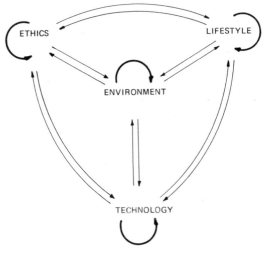

Fig. 11

Toward monumental environmental changes,[1] such as volcanic action, flood or drought effects, and earthquakes, man has tended to maintain a resigned posture. People live near volcanoes such as Vesuvius and Etna, though always in danger of being killed by a volcanic eruption. Following such a volcanic eruption, the lava often fairly rapidly becomes soil, and farming commences above the buried past. There are many examples of rivers which have shifted course suddenly due to natural geological factors. Mountains and islands both form and disintegrate generally over long periods of time but occasionally such changes in landscape occur quite rapidly. For example, Graham Island near Sicily built up from a 120m depth to a height of 60m above sea level, within a month during 1831. All that now remains of this island is a shoal. Camiguin Island in the Phillipines added over 600m to its height during 1871–5. In 1883 Krakatau Island in Indonesia was two-thirds blown away by sudden volcanic action.

More recently, two reputable astronomers[2] have predicted a number of catastrophic environmental changes for the early 1980s, resulting from the alignment of all nine planets of our solar system. The environmental effects they predict range from alterations in wind direction and velocity to changes in rainfall patterns and increased earthquake activity. In particular, they predict that much of southern California along the San Andreas Fault will be devastated at that time, and this will result in thousands of deaths and the almost total destruction of the city of Los Angeles.

But these sudden and dramatic environmental changes are not the only spontaneous environmental developments. In fact, such changes are generally of less significance than the more gradual and less notable environmental alterations which seem to have less impact in the minds of people. Such long-term environmental changes can be considered under headings of geological, evolutionary, and climatic. These changes have great historical significance, and are still operable today.

Long-term geological changes include factors ranging from the gradual drift of continents,[3] the erosion of mountains and hills and the slow changes in soil composition, to the slow evolution of rock formations. Such changes are always ongoing and indicate a maturation process for the environment. Such factors, however, only become relevant when a sufficiently long time span is considered, a time span made pointless by the crisis nature of current environmental problems.

An interesting exception to this relatively slow geological change, which has considerable interest in southern Queensland and northern New South Wales, would be beach and foreshore erosion and restoration. According to factors such as direction and intensity of winds, storm conditions, and wave directions, beaches build up and disintegrate over time in a complex fashion, only now coming to be at least partially understood.[4] Beach erosion following sand mining, and the attempts to restore foredune areas, must be seen within the context of the naturally changing beach environment, and not within any imagined constant.

Evolutionary changes apply to all forms of life within man's environment, including man himself. On an extended time scale, the evolution of various plants and animals has occurred and has altered the environmental radically.

Plant communities also undergo a form of evolutionary change towards a climax or most stable and permanent combination. Following disruption by man of a plant community such as a tropical rainforest, there spontaneously evolves a process of change in the community, normally back toward the original climax condition. Unfortunately, such spontaneous alterations of plant communities require a very long time, and if subject to other environmental changes such as soil erosion, or changes in animal life, the community may develop towards quite a different climax.[5]

The evolution of pathogenic organisms is still one of the most serious threats to man's survival. It was such a newly evolved organism which killed one-third of the population of Ireland in 1202-4, or which brought the Black Death which halved the population of Great Britain during the fourteenth century. A similar plague in

London in 1665 killed sixty-eight thousand people. In this century, a new type of flu developed just after World War I, and is thought to have caused more deaths (25 million) than occurred on all sides during the war.[6]

A third type of spontaneous environmental change is climatic. Many of the surface features of northern regions of America, Europe, and Asia were determined by the advance and retreat of masses of ice, brought about by relatively small changes in temperature and precipitation. It has been estimated that if all the ice masses were to melt, the oceans would rise by 60m, enough to inundate most major cities of the world. The long-term rate of rise of the oceans has been estimated to be 13cm per century, though just between 1930 and 1948 the level rose by 7cm.

But many climatologists are already suggesting that the retreat of the ice masses and the rise of the oceans is being reversed. The earth's mean temperature has dropped 0.3°C since 1940. This is enough to start the ice masses growing. Already it appears that a permanent ice link is forming between Greenland and Iceland.

Professor H. Lamb, director of the climatic research unit at the University of East Anglia, sees "Australia's abnormal rainfall, the icing up of Greenland and Iceland, the southward expansion of the Sahara, and the harsh dry spells in South Asia and parts of Central and South America ... as linked in the alarming [climatic] change".[7] He dates this shift in rainfall from about 1940. Dr. R. Bryson, head of the Institute of Environmental Studies at the University of Wisconsin, points out that climate can change greatly, due to relatively minor changes in other environmental factors.[8] From his observations, Bryson forecasts mass starvation resulting from what he believes is a drastic shift in world climate. Already, in fact, these climatic changes may have displaced up to 15 million people and caused famines in large parts of north Africa. A United Nations report[9] warns that up to 500 million people are threatened with starvation over the next decade, at least partially because of climatic changes.

While not all climatologists agree with the doomsday warnings of an approaching ice age, as well as more floods, deserts, and ensuing starvation, there is widespread agreement that the world climate is changing, and that the relatively bland climate of the early part of this century has come to an end.[10] Such climatic changes are complicated because the spontaneous changes are also affected by the man-made factors such as were discussed in chapter 10.

Self-induced environmental changes are of such a degree, and generally over such periods of time, that they are difficult to exactly predict or understand. The environment is never constant, being always in a state of flux. It is onto this changing environment that the effects of the other subsystems are imposed. And the resultant direction of environmental change is seldom simple or clear. For example, there are scientific opinions that increased air pollution will lead to higher temperatures (greenhouse effect) as well as other scientific opinions saying it will lead to lower temperatures (ice age effects).[11] These trends are to some extent offsetting and, in conjunction with naturally occurring environmental shifts, rather unpredictable.

SELF-INDUCED CHANGES IN LIFESTYLE AND ETHICS

As was discussed in chapter 4 both lifestyle and ethics are aspects of culture. The two have been discussed as separate subsystems because that has been the traditional treatment accorded by other writers who propose environmental solutions (see chapter 3). The essence of the model presented in this book however is that they are interdependent and solutions to environmental problems will only be effective if this and the other interdependencies between all subsystems are taken into account. When we come to discuss self-induced changes in both lifestyle and in ethics, this division could be maintained, but to combine the two, in terms of a discussion of theories of social evolution, seems more productive. Several examples will first be provided, pointing out such changes in lifestyle and ethics individually, then the argument will proceed to the more fundamental question of social stability, evolution, and revolution.

Lifestyles, or styles of everyday living, obviously change over time. Change is, of course, implicit in the term style. Previously, in chapters 5, 6, and 7, various influential factors with regard to lifestyle changes were discussed. But there may also be ways in which lifestyles can change in a manner which is self-induced, or at least self-sustaining.

For example, people such as Reich,[12] Roszak,[13] and even Yablonsky[14] suggest various trends in society which led to, or at least partially explain, the counterculture phenomenon in terms of causal factors within society. Reich, in particular, then goes on to delineate a number of common ideas or ideals held by such people, as well as various common factors of lifestyle, such as permissive sexual behaviour, long hair, gay clothes, and a weak commitment to a work ethic. The extent to which there is, or ever was, a true counterculture consciousness in the way suggested by Reich's "consciousness III" is open to debate. But the point is that while only a few people at first adopted the changed lifestyle, for the suggested external reasons, these changes have spread far wider, with a self-sustaining momentum quite separate from those reasons.

Another culturally significant change in lifestyle which may be subject to the same sort of momentum concerns sexual behaviour. A number of empirical studies[15] suggest that sexual activity now occurs more frequently outside marriage, that people have more sexual contacts, and that sexual attitudes are gradually altering from a reproductive to a recreational basis. Various possible reasons for this change, such as the availability of improved birth control technology, have been suggested. But the fact that such changes in sexual activities and attitudes have occurred throughout history[16] suggests that, at least to some extent, a self-sustaining style change is involved. In other words, some people may well adopt a new lifestyle with regard to sexual activity as a consequence of some external stimulus. This style then gradually diffuses throughout society. Perceptions that "everyone's doing it", or "it's okay as long as we like each other", may simply replace earlier perceptions that, "no decent girl does", or "every man wants to marry a virgin", and such perceptions, however true or false empirically, may help account for the change in lifestyle. Thus, while a lifestyle change may have had some causal factors, the change may develop a self-sustaining momentum which appears to be spontaneous to any individual or even to any community in which the original causal factors did not operate.

The same sort of style factor could be argued to apply to some changes in ethical values. That certain values are seen as in style, or trendy, helps people to agree with them.

It is easier to hold certain beliefs when they agree with the beliefs of others around us. We have all experienced the sensation of expressing an ethical belief in a group, and realizing that others do not agree with us. The tendency is either to change groups or to change values. To the extent that the latter happens, it is an ethical change devoid of outside cause. The likelihood that a person will change a stated ethical value in response to group pressure depends on factors such as age, sex, familiarity with others in the group, and personal history of the individual. The extent to which a stated ethical change is subsequently adopted by the person (conversion) is somewhat less, unless group pressure continues.

Self-induced changes in ethics may also be due to supernatural phenomena. For example, Christianity may be seen as having an inherent change factor. Many fundamentalist Christians believe that God determines everything that happens and that a close reading of the bible will reveal this predetermination to man. Currently in vogue moral values are all dismissed as part of God's plan, to perhaps tempt man, or as the workings of the devil, but the ethical changes are seen as predetermined.

Similarly, concepts of natural law imply that there are certain fundamental truths. If, for any of the reasons discussed in earlier chapters, ethics stray from those natural truths, there is a tendency back towards the truth. Plato argued that good and bad, right and wrong were fixed, and that man would, through knowledge, aim towards these good and right values. There would exist, at any one time, spontaneous movement towards the good and the right. That Utopia has not been attained only indicates the strength of countervailing forces.

Perhaps of more significance than these style factors are ideas of stability or change as the natural state of society. To what extent does a culture, though in a dynamic process, always aim towards equilibrium? To what extent is a culture involved in an ongoing change or evolutionary process?

In the nineteenth century, following the biological evolution theories of Darwin, various theories of cultural evolution gained popularity. There was generally contained a notion of cultural superiority and inferiority, and implied an evolutionary progress along a continuum toward a better, or higher type of culture. Belief in a steady evolution of human society became common among leading social scientists.[17] Marxism can, to a large extent, be seen as one such theory of cultural evolution.

Early in this century, such cultural evolution theories came to be greatly discounted by social scientists. Part of this rejection centred around perceived abuses of the cultural evolution theories. Such reactionary branches of cultural evolution theory assumed (1) that all changes in a particular culture were a form of progress (2) that all cultures over time go through the same stages of evolution or progress, and (3) that the less advanced cultures are similar to previous stages of more advanced societies.[18]

The fact that social Darwinism was often used or abused as a form of pseudoscientific justification for racism and other forms of paternalistic exploitation meant that progressives who rejected racism often also rejected evolutionism. According to this school of thought, all cultures were basically of the same value. Humanity was not a procession of various peoples marching toward the same goals, but was seen as being divided into many discontinuous cultures, each of which exemplified the variability of human inventiveness, and each being equally important.

Several authors[19] have recently reasserted the idea of cultural evolution. They argue that the gradual changes implied in evolution are the predominant social processes at any time, and that social stability and equilibrium are abnormal breaks in this natural flow of social change. Revolution they see as merely a marked acceleration of the evolutionary process.

According to this concept of social change and stability there are many influential factors impinging upon a culture at any one time. These may tend to provoke changes in the culture. But of more significance, the culture is itself in a state of change, an evolutionary state in which social change is spontaneous and directional. This evolutionary process will work at different speeds within different cultures, and often with widely divergent short-term effects. But the most important factor is that any culture is in a state of change at most times, that such change is a natural state, not an aberration, and that to understand a culture, this dynamic process must be recognized.

This same parade or train concept of evolutionary progress and change is implicit in most futurological studies. Kahn and Bruce-Briggs[20] of the Hudson Institute describe what they call "the long-term multifold trend of Western culture". Included in this are fifteen factors, such as increasing affluence, increasing literacy, increasing sensate concepts, and increasing concentration of economic and political power. These trends in an evolutionary fashion determine man's future. To predict the future, they basically project forward these historical trends on the assumption that "in most countries of the world, it looks as if this multifold trend is going to continue ... The multifold trend is conceived of as a long term base line which may include within it certain short term fluctuations, some of which may temporarily reverse the trend and slow the general forward movement in the next fifteen years." But, they argue, the long-term trends of social evolution will still continue.

Thus, for the purpose of this book, we can see change within both lifestyle and the ethical subsystems as either a self-sustaining movement, which was originally causally explainable, or as part of a more profound process of evolution. Given either set of assumptions, or both, which seems more likely, changes in these two subsystems can be accounted for without having to search for influences resulting from the other subsystems. While change may result from forces emanating from these other subsystems, the change may also result from internal forces.

SELF-INDUCED CHANGES IN TECHNOLOGY

From the discussion in other parts of this book, as well as from everyday observations, it would seem that the technological subsystem is the most dynamic aspect of the heuristic model. But all the examples of technological change so far discussed were attributed to causal factors external to the technological subsystem. There are also many examples, however, of self-induced technological changes resulting from factors contained within the technology.

Toffler[21] argues that "the reason for this [self-induced change] is that technology feeds on itself. Technology makes more technology possible, as we can see if we look for a moment at the process of innovation. Technological innovation consists of three stages, linked together in a self-reinforcing cycle. First, there is the creative, feasible idea; second, its practical application; and third, its diffusion through society. The process is complete, the loop closed, when the diffusion of technology embodying the new idea, in turn, helps generate new creative ideas."

Historically, we have many examples of technological changes which resulted more or less spontaneously from other technological changes. The development of the horse collar has been previously referred to (in chapter 6). This technology increased by perhaps as much as 100 per cent the power available to medieval agriculturalists.[22] Heavy, iron, mouldboard ploughs which required the strength of horses, as well as being able to withstand greater strains, increased in refinement and use, as a follow-up in technological change. Wagons as a technology of land transport greatly changed as a result of the new horse harness. The development of large, four-wheeled wagons with pivoted front axles, adequate brakes, and whipple trees, all resulted from the availability of horse power—itself unleashed by the technology of the horse collar. The development of relatively inexpensive land transport led to further changes in agricultural technology such as alterations in farm types and even the sort of crops grown, since it became possible to transport farm produce for longer distances.

One modern area of spontaneous technological change which is often quoted by technocratic apologists concerns space research and exploration. Criticism of the astronomical costs of the United States space programme has often been countered by the argument that space research has provided many practical benefits, i.e., the development of technology to explore space, for whatever reason, has led to technological changes in many other areas. Though the parameters seem unclear, it has been claimed by one American astronaut that there are 750,000 technological inventions and innovations which came into being through space research. A few examples of these new technologically derived consumer goods are aluminium blankets which weigh only 75g, an anti-fog compound which car manufacturers now use in windscreens, miniature ball bearings now used in dentists' drills, digital clocks, and the dehydrated food used by many boy scouts.

Space exploration, for whatever reason it was undertaken, has led to other technological developments. This technology then leads on to developments in other areas in a sort of diffusion pattern. In spite of the fact that this is a rather expensive way to devise better dentist's drills and food for boy scouts, this is a prime example of technological change without a causal factor external to the technology. There is a tendency for the same process to apply to any area experiencing technological change. When the technology is subject to rapid development as in space research, there are increased opportunities for such self-induced technological developments in a spin-off effect.

Another very good example of self-induced technological development centres around transistors. The transistor was invented in 1948, though it did not enter into practical use until 1953.[23] Because of decreased weight, vastly decreased power requirements, and similarly large increases in speed, the transistor led to innumerable changes in other forms of technology ranging from computers to space exploration, to the diffusion of propaganda and news.

Computers, already referred to as being dependent on the development of transistors, in turn change and will continue to change other aspects of technology. The technologies of mass communication, education, agriculture, weather forecasting, and even the means of distribution and personal selection of, and payment for, consumer goods may all be altered radically by spontaneous developments resulting from computer technology.

There are, of course, an infinite variety of such technological developments. For some reason, one aspect of technology changes, then other aspects are either caused to or allowed to change. This is the "closing of the circle" of technological diffusion to which Toffler referred.

The technological subsystem has already been described as the most dynamic of the subsystems posited in this heuristic model. We are now perhaps able to better understand why. While spontaneous change is possible in each subsystem, only with regard to technology is the change of an exponential nature. Development of one aspect of technology, for example the transistor, leads on to other areas such as space research, changes in communication patterns, computer design, etc. Then each of these areas, in turn, is capable of a multitude of spin-off effects.

This exponential rate of technological change, due to spontaneous technical growth, is the underlying factor in Toffler's discussion of *Future Shock*. These changes come at us at an ever increasing rate, and each technological change in turn spawns even more change. It is this inherent instability within the technological subsystem which contributes most to the instability of the overall system. This, then, accounts for much of the dynamic nature of the heuristic model, as will be discussed in more detail in chapter 12.

CONCLUSION

In this chapter we have examined the question of self-induced change within each of the proposed subsystems of the model. The environmental subsystem was discussed in terms of ongoing change at several levels. There are the long-term geological, evolutionary, and climatic changes which over time radically alter the environment. There are also more short-term changes, such as perhaps the marked climatic changes which may be now confronting us, and which threaten us with increased problems of drought, flooding, and human suffering. Finally, there are short-term spontaneous environmental changes such as earthquakes, volcanic eruptions, and erosion (in some cases).

Spontaneous changes in lifestyle and ethics were discussed on two levels. On the first level examples were provided of what could be called style changes which affect both how we live and how we think. Some people adapt their culture in some way, due to some reasons, and thus form a subculture. There may then be a tendency for the new lifestyle or ideas to spread, altering the culture of other people, even though the original causal factor may be inoperable.

Of more significance, perhaps, is the idea of culture being normally in an evolutionary state, and of stability being abnormal. According to this view, any culture is subject to a normal, directional process of change—of becoming. Professor Wertheim, on whose book much of the current argument was based, sees the direction of the cultural evolution as towards emancipation—from nature and from fellow man. But other cultural evolutionists see different end points for this teleological process.

Technology was seen to be subject to marked spontaneous changes, through diffusion of new ideas and methods. This was seen to have applied historically to factors such as the horse collar. Today this tendency toward spontaneous technological developments, particularly in areas such as electronics, medicine, and space research, presents an exponential rate of change, and is the most destabilizing subsystem of the model under discussion. This observation has been the subject of much social comment.

Such spontaneous changes in environment, lifestyle, ethics, or technology obviously tend to move the subsystems toward instability or disequilibrium. But the extent of this destabilizing influence is different for each subsystem, and will vary for each subsystem over time, though the technological subsystem tends to be the most unstable element, and contributes the most towards system instability. Environment, lifestyle, and ethics all tend towards stability while technology tends towards instability.

12 The heuristic model — Review and observations

In chapter 4 a heuristic model was introduced with the intention of facilitating the development, understanding, and analysis of various solutions (chapter 3) to the environmental problems which were discussed in chapter 2. In chapters 5–11, I explained in detail the nature of the relationships between the various subsystems in this model, as well as the tendencies towards stability or equilibrium within each subsystem and between each pair of subsystems.

In this chapter, the complete system will be reexamined, and the conclusions about stability and instability, drawn from an analysis of relationships between subsystems, will be looked at in terms of their meaning for the total system.

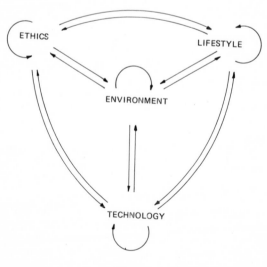

Fig. 12

STABILITY AND INSTABILITY OF THE OVERALL SYSTEM

In chapters 5–10 it was found that equilibrating forces existed between each of the subsystems, so that a change in one would be reflected in changes within one or more of the other subsystems. Also, it was observed that frequently, as one subsystem changed, countervailing forces were set up by another subsystem which tended to oppose or offset the changes in the first subsystem.

We have found that any change in a subsystem is responded to by other changes towards agreement—and by pressures against any further changes. On this basis system stability would be the obvious, eventual result, although such stability would not be static—but a dynamic equilibrium. Nevertheless, it would be around a fixed point that the overall system would fluctuate. Long-term changes would be precluded.

But this form of reasoning assumes that the various subsystems are themselves capable of some form of optimizational behaviour which is consistent with attainment of an optimal state for the complete system. Experience from natural systems indicates that such a state of affairs is highly unlikely. In fact, "If particular subsystems of a larger system operate so as to optimize their own individual 'good' [performance indices], the net result will almost never be overall system optimization."[1] What happens in a natural system such as the one being discussed here is that the various subsystems do tend toward optimal states. But such tendencies are tempered not only by pressures from the other subsystems, but by restraining pressures from the complete system.

All of the tendencies and pressures so far discussed point towards the heuristic system described here, as moving toward a form of stability. All we have studied then is what exactly determines the conditions of the stable state. We see that while equilibrating tendencies within and between the various subsystems are pertinent, these occur within a set of constraints imposed by the complete system. Long-term changes of an evolutionary nature cannot be explained by such an analysis.

But it is obvious that changes of more than a fluctuation around the baseline nature do occur. The explanation of these changes is largely contained within chapter 11. There, self-induced changes within each of the subsystems were discussed. These changes, of course, tend to have a destabilizing influence in that they require a corresponding change both within the subsystem and from other subsystems to maintain an equilibrium situation. Likewise, a shift in the overall system may result from such destabilizing tendencies of subsystems.

When assumptions of cultural change and natural evolution are accepted, overall system instability is suggested. But far more important is the inherent instability of the technological subsystem. In chapter 11 we discussed how technological changes tend to build upon previous technology, and in turn spark off further changes in a mushrooming fashion. Thus a change in the technological subsystem often produces still further changes, resulting in inherent instability. It is consistent with general systems theory that an inherently unstable subsystem can lead to overall system instability, and thus to instability within the other subsystems.

Within the complete system there are stabilizing influences and destabilizing influences. The environmental crisis can be interpreted as partially a victory for certain of the forces toward system instability, and this instability moves the system in directions of even further instability. Rapid human population growth, the widespread use of mono-culture farming systems, pollution levels within natural areas reaching threshhold-of-complete-breakdown conditions, and the social unrest encouraged by such factors and others, all indicate system instability.

HOW TO CHANGE THE SYSTEM—ENVIRONMENTAL SOLUTIONS

The reader might, at this stage, agree that the model is useful for conceptualizing current problems but conclude that it does not suggest solutions for the environmental crisis. Indeed, it may even suggest that solutions are precluded. The system has a dynamism of its own resulting in us being locked into an ever worsening environmental crisis situation, from which there is no escape save catastrophe.

But such a gloomy conclusion is only based upon a cursory examination of the interrelatedness of the subsystems as outlined in chapters 5–10. These are taken as indicating that no change is possible. Unfortunately, a similarly naive understanding of chapter 11 indicates the opposite, that change is endemic within the overall system. Ever increasing disequilibrium between subsystems is seen as leading toward some sort of ever more widely oscillating effect— and eventually toward a complete breakdown. It is my contention that neither gloomy conclusion necessarily follows from the model. Change is possible, but not necessarily endemic. Mankind might possibly be locked into a course toward self-destruction, but such is not suggested by this model.

While the system has tendencies toward stability and instability, these do not preclude the application of rational human choice. Man still has the option of rational intervention to alter his destiny. An assumption of this model is not the preclusion of intervention, but that intervention to ensure desired changes is possible. Specifically what the model does suggest is the way that environmental solutions must be applied if there is to be any hope of alleviating the crisis situation. It provides a framework for planning strategies. If a strategy is planned within the context of this model, it might well still fail, however, because the environmental problems are already too great for the type of strategy employed, or because the strategy was not employed with enough pressures in the right directions.

From the operations of the model, we can observe that the type of environmental solutions with the most chance of being effective are those that—(a) exploit or utilize the inherently unstable aspects of the subsystems; and—(b) utilize as many subsystems as possible, influencing or altering them in a given, mutually supportive direction. In other words, change is attempted wherever it is easiest to effect, i.e., at the least stable point, and pressure is applied at as many points as possible, so that mutually supporting change is achieved. The following are several examples of just what is meant by this approach.

Utilizing the inherent instability of the system is exemplified by, for example, technological developments which facilitate the technology of controlling environmental destruction, as were discussed in chapter 10. This technology may be an unintentional result of developments within the technological subsystem, in a sort of spin-off effect, as described in chapter 11. Similarly, spontaneous changes within the environment may help solve other environmental problems. If climatologists' warnings of an approaching ice age, with falling temperatures, are correct, then ecologists' warnings of a greenhouse effect from air pollution (i.e., rising temperatures) partially offset them. That is, a self-induced change within a subsystem may provide, or contribute to, some form of solution. Also, as we found in chapter 11, there are self-induced or self-sustaining changes possible in both lifestyle and ethics. By encouraging those changes which are consistent with ecological reality, environmental solutions may be facilitated.

Each of these ideas, based on the inherent instability of the subsystems, does hold promise of moving the system toward a resolution of the environmental crisis. But the results will be minimized unless pressures are applied to each subsystem. A multi-pronged attack, in other words, holds the greatest hope for constructive change. While it is helpful to develop technological tools to control environmental destruction, and it is also helpful to encourage social change toward more frugal lifestyles, the greatest impact is achieved when all changes are mutually supporting. It can be seen that many of the proposals described in chapter 3 to solve the environmental crisis are more mutually supportive than competitive. There is no question of either/or, but a question of which combination of tactics will be the most supportive of moving the overall system towards a more stable state, that is, a state wherein the environmental crisis problems discussed in chapter 2 are under control.

Using terms like *constructive change* obviously introduces heavily value-laden premises. But why not? It is a misunderstanding of the nature of systems to assume a goal-free or directionless existence. Integrated systems are goal-oriented systems.[2] It is possible to utilize value-laden terms like the *purpose* of a system, and to evaluate potential outcomes as good or bad. System optimization is by definition the goal of an integrated system. Since the very criteria for such judgments are integrated within the system, it is possible to describe a system in terms of ends and goals.

Also, it is implicit from the development of the model that a more stable state is a goal contained within that system. That the current environmental crisis threatens the overall system seems clear. That a goal of the system is its own survival also seems clear. That one can objectively speak of a system having a goal of decreased instability, when that instability threatens the system, also seems consistent with systems theory.[3] Therefore, we can speak logically of this system itself having an implicit goal in increased stability, albeit of a dynamic nature, and any activity consistent with the attainment of such a goal should be utilizable by that system.

In other words, a dynamic, natural system, such as is posited within the heuristic model, contains within it factors both constructive and destructive to the long-term viability of the system itself. The continued existence of a natural system is a goal of that system and, therefore, if instability threatens the continuance of a natural system, the limitation of that instability becomes the purpose of changes within the system. It is not reification to refer to a goal or purpose of a natural system, such as that under discussion. The goal of system survival is implicit in that system.

CONCLUSION

In this chapter we have looked once again at the overall system which was introduced in chapter 4, and developed and explained in each following chapter. The system indicates neither inability to change nor endemic change. There are aspects within the system, however, which are relatively stable, and other aspects which are relatively unstable.

The most likely successful mode of attack on the environmental crisis was found to be of a twofold nature. It utilized or exploited areas of inherent change within the system, and it was one which exerted influence or pressure in as many areas (subsystems) as possible. Such a strategy, where the pressures are supportive, would have the most chance of moving the system in the desired direction. This conclusion is, of course, fully consistent with systems theory.

The information feedback between subsystems, which normally minimizes overall systems change, is reduced by such a strategy. The information feedback actually comes to support change in the overall system and to sustain and institutionalize changes which have occurred.

In the following chapter (chapter 13) a concrete example of what is meant by such a multi-pronged attack will be presented and discussed, within the theoretical discussion from this chapter. Utilization of inherent instability, pressure for change on many fronts, and the mutually supportive information feedback system thus engendered are the essential elements utilized in the following specific application of this heuristic model.

PART 3

Application of the model

In Part 3 the heuristic model, which has been under detailed discussion, is reviewed as a complete entity. Theoretical observations are made, concerning the tendencies towards stability and instability of this model. From this, conclusions are drawn on how best one might implement changes within the natural system which this model represents.

This book concludes with the application of this model to the currently crucial environmental problem of mass starvation. Observations are made concerning the chances of success of suggested solutions to these problems, and it is argued that a multi-pronged attack, as indicated by this model, will have the greatest chance of success. The problem of mass starvation is an example used to illustrate the use of this model to provide a framework within which solutions might be formulated.

13 Application and utilization of the heuristic model

Already in the examples used in discussing the development of this model, many applications have been suggested. Numerous aspects of the environmental crisis have been discussed, and simple cause-effect solutions based on linear thinking have typically been found wanting. Frequently it was found that just as the environmental problems are part of a complex system or world ecosystem, so too are the ways to solve or alleviate those problems. In this chapter the utilization of the proposed heuristic model will be discussed, both in general terms and with a concrete example drawn from chapter 2.

Firstly, when discussing the relevance or usefulness of a model, it is important that the purpose of the model be borne in mind. A simulation model attempts to mime objective reality and to be predictive of future behaviour. Thus the usefulness of a simulation model is dependent on the accuracy with which this predictive role is accomplished. Such accuracy can be tested by utilizing information not already used in deriving the model, or future predictions of such behaviour can be made and then compared to the actual behaviour as it occurs. But in either case the usefulness of the model is, at least in theory, open to tests of objective mathematical reliability.

A heuristic model, on the other hand, does not attempt to be predictive in the same way (see chapter 4). Since it is not meant to be quantifiable, the strict mathematical testing of a heuristic model is precluded. Rather, the usefulness of such a model must rest on other, less objective, criteria, like the degree to which it provides a mental set or perspective to aid understanding, and there is no immediate way to test the extent to which it does this. Nevertheless, one aspect of the environmental crisis has been selected, and discussed in detail, utilizing the heuristic model of this book as guide.

In chapter 2, after discussing the environmental problems of population, pollution, and resources, the more specific human problem of starvation was mentioned. This, it was suggested, is both a result of environmental problems, and an environmental problem in its own right. Chapter 2 concluded by arguing that the interrelated nature of the environmental problems, and the crucial factor of time, allow the term *environmental crisis* to be used in a reasonably objective fashion. And this environmental crisis is nowhere more apparent or more pressing than in regard to food supply and starvation on a global basis.

In chapter 2 we saw estimates of from 10 million to 20 million people starving to death annually in the recent past. Current estimates of 50 million to die from starvation within the next twelve months, and up to 500 million within the next decade, were also discussed. As well, there are estimates indicating that one-quarter to one-third of the entire world's population (50–60 per cent in underdeveloped countries) were suffering from malnutrition. Current reports from a number of sources indicate that these problems are becoming rapidly worse in many parts of the world, but particularly in equatorial Africa and the Indian subcontinent. It appears that we are in a rapidly deteriorating situation with regard to the world's food supply and starvation.

This chapter will look at the problems of starvation and malnutrition, in terms of the usually suggested approaches, as well as in terms of the model developed in this book. Is mass starvation an environmental problem? How do the usually suggested solutions fare? What are logical short-term and long-term solutions? How can this heuristic model be used as a conceptual framework to rationalize solutions to this acute problem?

FOOD SUPPLY AND STARVATION AS ENVIRONMENTAL PROBLEMS

The crisis situation in the world with regard to insufficient food, of inadequate quality, and the resulting malnutrition and mass starvation is at the very crux of the environmental crisis. This assertion is based on the following factors.

a. Almost by definition the environmental crisis is a crisis concerning man within his environment. If there was no man, there would be no crisis, or at least no one to define it as a crisis. Because what we are looking at is man in the environment, and interacting with his environment, i.e., human ecology, we can base a judgment on observations either of man or of the environment. It is obvious that mass starvation, such as currently exists, indicates a serious problem in this man-environment relationship. The fact that people are starving to death is an environmental problem, because it represents a fundamental breakdown in the human ecology system.

b. Extreme poverty, malnutrition, and starvation all act to preclude fundamental solutions to environmental problems—even the problem of starvation! Where a drop in crop production could mean starvation, it is difficult to get a farmer to try a new seed variety or farming technique. While people are worried about finding enough to eat today or this week, it is difficult to raise awareness of longer-term environmental problems such as pollution, soil erosion, resource depletion, or overpopulation.

c. Attempts to solve or alleviate problems of starvation often result in a worsening of other environmental problems. Children, for example, are often desired by peasants because of the help they can provide in the fields. Also, in a very insecure situation, one's only hope for old age security may be that one's children will provide. Yet this solution, while perhaps logical for an individual, is disastrous for a nation. Likewise, agricultural chemicals are used with the only criteria being immediate changes in crop yields. The effects of one person's activities on other producers, the pollution effects in the waterways, and the long-term health factors are all ignored.

d. The final reason for asserting that the problem of mass starvation is an environmental crisis is an ethical one. As long as people anywhere in the world are dying from starvation, then there is a problem. The basic tenet of this book has been systems thinking. We cannot dissociate ourselves from problems in other parts of the world since the world is a global system.

It is one thing for conservationists to worry about protection of foredune areas of beaches, saving urban parkland, and the protection of endangered species of wildlife; while not wishing to denigrate the importance of these environmental concerns it would seem to me that prevention of human deaths is more important. When there is a massive breakdown in human ecology, such as we now face, there can be no doubt that this is an environmental crisis of serious proportions.

REDUCING STARVATION AND MALNUTRITION

There are three methods suggested and occasionally used in attempting to reduce starvation on a global basis. These are—(*a*) better utilization (less waste) of food supplies;—(*b*) production of more food; and—(*c*) provision of less people to feed and, therefore, presumably, resulting in less starvation. A short discussion of each of these suggestions will follow.

a. Better utilization of food

Arguments concerning the better utilization of food supplies generally take the following form. In a country or region, a certain amount of food is annually produced (or could be produced), but only a certain percentage of this ever reaches the consumer, while the rest is eaten by vermin, rots, or is ruined by some disease. The argument, therefore, concludes that if we could only prevent this loss, many more people could be fed. Fungal diseases alone yearly destroy enough food to feed 300 million people.[1] Altogether the cereal rusts cause an estimated annual loss of $500 million. Such an amount of grain, if it could be saved, would help keep the 2 billion hungry people of the world from having to subsist on a starvation diet. But even when food has been produced and protected from disease and vermin, problems of transportation, distribution, and simply bureaucratic bungling may prevent its being utilized to prevent starvation.

While there are doubtless many ways in which food losses could be reduced, and emergency food supplies more efficiently handled, there are obvious limits. It does not seem possible to totally eradicate pests, crop diseases, human greed, and bureaucratic bungling, though their detrimental effects might be curtailed by more intense effort. There is certainly room for reduction in starvation by the better utilization of food, but to think that this is more than a stop-gap measure is to be unrealistically naive.

b. Production of more food

The most obvious solution to shortages of food and to problems of starvation would seem to be to produce more food. Much of the literature on problems of starvation seems to regard this as so obvious, in fact, that it is quite implicitly assumed, and is rarely examined, or compared to other approaches. It is such simplistic thinking, of course, which this book means to attack. Environmental problems and environmental solutions must be seen in a systems framework. There is little scope for simple, linear approaches to solving complex ecological problems.

In the past, agricultural production of food has increased in two directions: the areas farmed have increased and, simultaneously, changes in technology over the long term have increased yields per unit of land. The changing agricultural technology has, in turn, occurred in two ways. The first is changing methods of preparing the soil, harvesting the crops, rotating crops, etc. The second factor refers to changing agricultural inputs, such as the use of inorganic fertilizers, pesticides, spray irrigation, hybrid seeds, etc.

While there are still ways in which agricultural production can be increased by bringing new areas under crop production, as well as by the use of better soil husbandry, the greatest scope for dramatic increase seems to be through increased agricultural inputs. The use of extra water and fertilizer, as well as newly developed short stem varieties of wheat, rice, and maize have led to examples of marked increases in agricultural production on test plots in many parts of the third world, and these production increases have been termed the Green Revolution.

Some people, such as futurologist Herman Kahn, predict "a world food surplus rather than shortage for the mid [nineteen] seventies. A moderate world agricultural depression due to overproduction is much more likely than a Malthusian famine caused by increased population outstripping agricultural land."[2] The enthusiasm of Kahn (and a few other super-optimists such as Colin Clark[3]) is not shared by even the people responsible for the Green Revolution, such as Nobel Prize winner Norman Borlaug. At best, such people see the Green Revolution as providing an invaluable, though only temporary, breathing space in the battle to stabilize human population numbers. One group of scientists writing in the now famous "Blueprint for Survival",[4] argues that "Whatever their virtues and faults, the new genetic hybrids (i.e., Green Revolution seed grains) are not intended to solve the world food problem, but only to give us time to devise more permanent and realistic solutions".

There are several main constraints to the Green Revolution. The first is the agricultural extension problem of getting peasants to experiment with new crop varieties. A family which is chronically on the verge of starvation may not be very anxious to experiment with new varieties, where failure could spell disaster. Also there are tremendous problems of shortages of the resources on which the Green Revolution depends. Fertilizer prices in some third world countries have tripled over the past few years, and phosphates on a world basis are already in short supply. Water, for irrigation, is in most places a limiting factor, and the almost universal overuse of water leads to problems of soil salination—often reaching a point where the soil is made useless for growing crops. The increased use of fertilizers and pesticides leads to increased pollution in the drainage water, often threatening aquatic life—and the important protein source, fish. It is a fundamental ecological concept, that the more an ecosystem is simplified (e.g., toward Green Revolution monoculture), the more vulnerable the crops are to attack by pests or disease, and the more catastrophic will be such an attack. The debilitating attack of blight in the United States corn belt during 1970 destroyed an estimated 710 million bushels, roughly 17 per cent of the anticipated crop.[5] Such an attack could be even more widespread, and would have tragic results in a country less able to combat such diseases. Finally, the Green Revolution has run up against cultural problems of human greed, corrupt officialdom, inadequate distribution systems for the inputs and for food produced, as well as an inadequate economic system, to allow potential consumers in underdeveloped countries to purchase the produce. While these latter problems are not technically insurmountable, they will require a great deal of time and effort on the part of the governments involved. So far there is little evidence that such effort is forthcoming. In fact, even worsening trends may be in evidence.[6]

While most critics of the Green Revolution would regard it as perhaps a necessary evil, few totally condemn it, conceding that it must be seen as a stop-gap measure in the environmental crisis. Although there are many harmful environmental results from the Green Revolution, in the short term there appears to be little choice but to accept these. The interrelated nature of the environmental problems must be recognized, however, as forming a barrier to the long-term success of any such technical fix to the problems of overpopulation and starvation.

The whole logic behind the Green Revolution is indicative of non-systems thinking. Inputs are increased (or attempted) with little thought of pollution problems, resource depletion or of social effects. What works wonders in a test plot it is hoped will likewise work wonders on a country-wide, or world-wide, scale. Within the heuristic model presented in this book, the pitfalls of the Green Revolution can be better understood. Hopefully, this model can also help indicate ways in which the advantages of the Green Revolution can be gained, without so many of the disadvantages.

While increased agricultural production has been the most commonly suggested way to increase food supplies, other suggestions are also made. Farming the oceans, or aquaculture, is frequently seen as a partial solution, particularly towards alleviation of protein shortages. Already, in fact, sea food provides almost one-fifth of the world's protein. While there are many ways by which increased protein can be won from the sea, by, for example, fish herding, or the harvesting of phytoplankton, there are also many problems. Most of the plans for increasing the yield from the sea completely disregard the effect of pollution, in a clear example of non-systems thinking.

There are also various suggestions for exotic synthetic food supplies, from petroleum, for example, though more commonly such schemes utilize plant material. By means of the Vepex process, for instance, nearly any green vegetative matter can be fractionated into a powder that is 45 per cent protein, low in fibre content, and digestible by humans. While technically this does appear promising, it would be most naive to see it as any serious long-term solution. Food consumption patterns are a social phenomenon, not amenable to rapid, radical alterations. There are many examples of Asian peasants starving because rice was not available, even though there was ample wheat. If there are such problems in people changing food habits, it is questionable what short-term success will be had by exotic synthetic food compounds. In the longer term, however, in conjunction with an active propaganda campaign such products may become acceptable. All that means is that social change must be encouraged to facilitate use of such food.

But any attempt to reduce starvation by increasing the food supply will obviously eventually fail if the numbers of people to be fed continue to increase. There is a finite limit to the sustainable long-term productivity of land and the seas. Regardless of whether you accept that this limit is several times above current production, or that we have already passed the limit, the fact remains that there is a limit. Human population numbers doubling every thirty-three years means that increasing food production is, at best, a stop-gap measure. At worst, increasing food production may even reduce the long-term sustainable food production capacity of the globe.

c. Reducing the number of people to be fed

The third way of reducing starvation is to reduce the numbers of people to be fed. This can be seen in either a literal sense of decreasing population, or in a relative sense, of decreasing the population growth rate. The most common arguments along this line recommend that birth control techniques be more encouraged and utilized to reduce births, reduce the rate of population growth, and perhaps eventually, due to natural deaths, lead to stable or slowly declining population numbers. This logic has led to several countries such as India and Japan adopting active programmes of encouraging birth control through propaganda as well as various economic incentives. Unfortunately, the results have often been rather unspectacular.

Various authors have recommended that developed countries encourage birth control in the underdeveloped countries as a condition of, and by means of, foreign aid. Several economists have calculated benefit/cost ratios on the use of foreign aid moneys to promote birth control and thus reduce population growth. The results show that it is far simpler and cheaper to prevent a birth than to provide for an extra person. Lipton in one study found that "the addition to income per head in a poor country caused by spending $100 on birth prevention is, at very least, fifteen times the addition to income per head from the best alternative use of $100".[7] Enke was more accurate than this. He calculated that "the amount of money spent each year on birth control can be 100 times more effective in raising income per head than the amount of money spent each year on traditional productive investments ... The benefit/cost ratio is 22 to 1 already by the fifth year and rises to 82 to 1 by year 30."[8]

While there are many questionable assumptions which must be utilized in producing such estimates, there can be no doubt of the efficiency of resources devoted to population control, even if the actual benefit/cost ratios are open to debate. The provision of propaganda and free or subsidized birth control methods as well as other economic inducements are designed to help overcome pronatalist social pressures, i.e., the social pressures which encourage people to have children.

Such relatively benign efforts to reduce population, while being no doubt in the interests of people in general, may be unwise for certain people in particular. In a society where one's only old age security is provided by one's family, policies which discourage a couple from ensuring that they have at least one son may, in fact, hinder their personal security. This once again is an example of linear, or non-systems, thinking. A solution such as birth control may be ineffective, or even in some cases harmful if not combined with many other, mutually supportive, changes.

More dramatic than encouraging birth control techniques are various suggestions for reducing population numbers through conscious decisions to let people starve. The logic behind this approach is that, as we cannot possibly keep everyone alive in a world of rapidly increasing population, a decision to maintain only some people, and to allow others to die now may, in the long run, cause the least starvation, and be the most humane. The most famous argument along this line has been presented by the Paddock brothers.[9] In military medicine, they explain, the term *triage* is utilized. This refers to a division of the wounded into three groups: those who will survive without medical care, those who will not survive even with medical care, and those who will survive only by means of medical care. The scarce medical resources are, therefore, logically devoted to the latter group, and in this way, deaths are minimized. Adapting this triage concept to a world food shortage situation, they recommend that hungry nations be divided into three classes on the following basis:

1. Nations in which the population growth trend has already passed the agricultural potential.
2. Nations which have the necessary resources and/or foreign exchange for the purchase of food from abroad and which therefore will be able to cope with their population growth.

3. Nations in which the imbalance between food and population is great but the degree of the imbalance is manageable.

Obviously, the authors recommend that food be given from donor nations only to countries of the third type. This plan for the use of food supplies, they argue, will reduce starvation in the long run, more than any other food-giving policy. This can be seen in some sense as a drastic form of population control. By intentionally not feeding certain people and allowing them to die, there will in the future be less people, and less starvation.

Most people who comment upon this triage scheme seem to reject it as being too cruel and inhumane. This criticism is, in my opinion, often based on a poor understanding of, or lack of knowledge of, their book. The Paddocks do not argue that triage is good or humane, but that in a crisis situation it is the least bad and the least inhumane of several nasty alternatives. When this book was published in 1967, the date when famines would become severe was estimated to be 1975. This prediction was roundly condemned at that time as being too pessimistic. The United States Department of Agriculture maintained that United States food supplies alone could feed the starving of the world at least until 1984.[10]

We can now see that the Paddocks were right, that serious famine is a problem already, and that United States food stocks are totally inadequate. Paddocks used systems thinking in their recommendations, seeing triage as part of a solution along with improving agricultural technology, promoting birth control, and foreign aid to encourage industrialization of certain areas, and policies to promote the rational distribution of food supplies within nations and to reduce food losses due to disease and waste. That is systems thinking quite in line with the tenets of this book.

APPLICATION OF THE HEURISTIC MODEL

The three most commonly suggested ways of solving or reducing the problems of starvation, i.e., reducing food losses, increasing food production, and reducing the number of people to be fed, have all been seen to basically use non-systems thinking. Whether you argue that food supplies should be increased, more carefully stored and distributed, or that steps should be taken to actually reduce the numbers of people to be fed, there can be little doubt that some improvements in the starvation problem will be achieved. But unfortunately the improvement may be very short-term, and in the end actually harmful. For example, there is no doubt that increased ap-·plications of commercial fertilizers will boost crop yields, increase food supplies, and thus reduce starvation. But if such agricultural technology (technological subsystem) is not judiciously applied, this increase in food production may be offset partially by other losses as, for example, decreased fish supplies due to water pollution from agricultural chemicals, or from soil deterioration (environmental subsystem). Also, if no attempts are made to curtail population growth, there may be an actual increase in starvation, if the overloaded food production system breaks down in any way. The increased use of monoculture cropping, of course, makes such a breakdown more likely.

Attempts to reduce human population numbers, either in absolute terms or in terms of a reduced growth rate, can be expected to reduce starvation problems. But such a goal, it has been found, is virtually impossible to achieve by means of the usual linear approaches. To propagandize or threaten people into reducing their birth rate (lifestyle subsystem) will have little effect if, for example, there is a strongly held ethical belief in the value of having large families, or if the economic system is such that one's only old age security is provided by the family. Similarly, such a reduction in birth rate may be unachievable within a society where woman's only role is to reproduce. Yet large family size may militate against any form of liberation of women, in a sort of vicious circle.

All of these ideas and observations indicate the need for systems thinking. This book has argued the interrelated nature of environmental problems. To ignore this ecological fact is to run headlong into linear thinking—and to encounter problems such as have been discussed.

In chapter 12, it was suggested that the heuristic model indicated that any viable environmental solution would be of a twofold nature. It would utilize or exploit areas of inherent instability or change within the system, and it would exert influence or pressure in as many areas (subsystems) as possible. In this chapter, we have seen that most approaches to trying to solve problems of starvation do not utilize this systems perspective, and therefore fall short of reaching a solution.

The Green Revolution is largely an attempt to utilize the inherent instability of the technological subsystem, and specifically agricultural technology. There are other ways in which technological developments could be promoted, as, for example, the use of solar energy to run irrigation pumps. This exploitation or utilization of the inherently changing nature of technology occasionally runs amok, when the repercussions of the new technology on the environmental subsystem are ignored. There are ways, of course, in which these effects can be modified, provided foresight and systems thinking has been employed. But any new technology is useless, unless utilized by people (lifestyle). This then is a social problem, but its resolution requires more than effective agricultural extension techniques, as is so often assumed. As long as there are factors within the lifestyle or the ethical subsystem which preclude changes in technology, then technological solutions will have little success.

Likewise, attempts to promote anti-natalist values in a culture, will have little effect unless birth control technology is available, and ethically acceptable. The lifestyle of the people must similarly be conducive to a reduction in the birth rate. The environment, if perceived as overpopulated, will also help promote birth control.

Attempts to reduce wastage of food will have greater success the more that modern technology of transport and storage is utilized. But such technology will only be developed, adapted, and implemented within an economic and social system conducive to such change. Ethical values in favour of the traditional ways might work against such changes. Ethical values of frugality might promote such changes. Also, there are very good environmental reasons for some wastage of food. To assume that a pesticide, for example, will be able to completely prevent insect or fungal infestation flies in the face of ecological reality. In fact, there are numerous studies which indicate that attempts to reduce food losses to pests, by the use of pesticides, may actually increase such losses.[11] That is, the most efficient way is to simply assume that some food will be eaten by pests, because to try and eliminate these losses may be counterproductive, and may well result in increased losses.

The application of this heuristic model to the problems of mass starvation indicated that there is a need for changes in each of the subsystems. All of the usually applied solutions, production of more food, prevention of food wastage, and prevention of births, can be usefully employed. But this model indicates that this application must be in a complementary fashion. While change is most likely in areas of inherent instability, such as the technological subsystem, comprehensive or meaningful changes are only possible where pressure has been applied to all subsystems in a complementary, multi-pronged effort. For example, changes in agricultural technology may be of little use, if the culture does not change enough to accept and utilize this technology. A desire for smaller family size will not be realized unless birth control technology of a suitable nature (in the eyes of the potential users) is available. In other words, a change within one of the subsystems can be largely offset by another subsystem, unless complementary changes have occurred.

CONCLUSION

The implications of this heuristic model for problems of mass starvation are clear and straightforward. Efforts toward a solution must utilize areas of inherent instability within the system, particularly the technological subsystem. Also, changes must be effected in as many subsystems as possible, and always, of course, in a complementary fashion. A multi-pronged attack, employing all the techniques discussed earlier in this chapter, will have the most chance of success. The heuristic model developed in this book helps explain how such a strategy would work.

While there is little doubt that such an approach could alleviate these problems, there is no implicit assumption that mass starvation as a problem can be solved. It may well be that human population numbers and resource depletion have progressed to a state where no absolute solution exists. It is not the intent of this book to argue that such a solution does exist. The intent is to provide an analytical framework within which a direction for solving environmental problems, such as starvation, can be logically formulated.

Having used this heuristic model, the directions in which efforts to solve such problems should be applied appear relatively clear. Unfortunately, this systemic nature of environmental problems is still frequently ignored in political reality, and linear thinking is employed to provide one-step answers. Such "answers", bound for failure, are precluded when this heuristic model is utilized as a frame of reference.

Notes
to
text

CHAPTER 1—THE IDEA OF AN ENVIRONMENTAL CRISIS

1. G. Phillips, "The Environment", *Australian*, 16 July 1974, p.17m
2. W. Henderson, "Community Must Help Bear Massive Costs", *Australian*, 16 July 1974, p.17.

CHAPTER 2—FACTORS IN THE ENVIRONMENTAL CRISIS

1. C. Stewart, "The Hazards of Population Forecasts", pp.559–62.
2. G. Taylor, *The Doomsday Boook*, p.214.
3. For example, R. Bryson, "A Perspective on Climatic Change", pp.753–60.
4. J. Maddox, *The Doomsday Syndrome*, pp.62–93.
5. President's Science Advisory Panel on the World Food Supply, *The World Food Problem* (Washington: Government Printing Office, 1967), p.2:5.
6. *United Nations in Action*, no.74/17 (24 May 1974), p.2.
7. "Heading to Food Crisis", *Courier Mail*, 25 October 1973, p.6.
8. J. Loraine, *Sex and the Population Crisis*, pp.163–75.

CHAPTER 3—A REVIEW OF THE PROPOSED SOLUTIONS TO THE ENVIRONMENTAL CRISIS

1. J. Platt, "What We Must Do", p.283.
2. Ibid., p.291.
3. I. Horowitz, "The Environmental Cleavage", p.129.
4. B. Skinner, *Beyond Freedom and Dignity*, p.2.
5. Ibid., p.12.
6. Ibid., p.79.
7. Ibid., p.173.
8. A.Spilhaus, "Ecolibrium", p.711.
9. R. Siu, "The Role of Technology in Creating the Environment Fifty Years Hence", p.97.
10. A. Berry, *The Next Ten Thousand Years*.
11. R. Fuller, "An Operating Manual for Spaceship Earth", p.370.
12. Ibid., p.374.
13. B. Commoner, *The Closing Circle*, pp.142–46.

14. Ibid., pp.88–89.
15. J. Maddox, *The Doomsday Syndrome.*
16. J. Maddox, "A Great Many of the More Scary Arguments of Ecological Catastrophe Have Been Shown to be Chimerical", p.311.
17. H. Coombs, *Matching Ecological and Economic Realities*, p.9.
18. Ibid., p.11–12.
19. W. Westman and R. Gifford, "Environmental Impact", pp.819–25.
20. For an excellent presentation of this argument see J. Passmore, *Man's Responsibility for Nature*, pp.3–40.
21. L. White, "The Historical Roots of Our Ecological Crisis", p.1,205.
22. E. Fiske, "The Link between Faith and Ecology", *New York Times*, 4 January 1970, Section 4, p.5.
23. H. Skolimowski, "Technology v. Nature, p.51.
24. E. Aarons, *Philosophy for an Exploding World.*
25. J. Wright McKinney, "The Individual in a New Industrial Age", p.6.
26. L. Orleans and R. Suttmier, "The Mao Ethic and Environmental Quality", p.1,173.
27. W. Metcalf, "The Ethic of Frugality", pp.30–32.
28. C. Birch, "People Are Dying because of the Air They Breathe", p.319.
29. B. Weisberg, "The Politics of Ecology".
30. P. Ehrlich, *The Population Bomb*, p.108.
31. P. Ehrlich and R. Harriman, *How to be a Survivor*, pp.124–26.
32. Ibid.
33. M. Bookchin, "Toward an Ecological Solution", p.14.
34. Ibid., p.15.

CHAPTER 4—A MODEL FOR UNDERSTANDING ENVIRONMENTAL PROBLEMS AND SOLUTIONS

1. P. Weiss, "The Basic Concept of Hierarchic Systems", p.31.
2. H. Buechner, "The Ecosystem Level of Organization", p.45.
3. For example, W. Sewell, "Environmental Perceptions and Attitudes of Engineers and Public Health Officials", pp.55–58.
4. "A Study of Vasectomy in Brisbane", Department of Social and Preventive Medicine, University of Queensland, Project 328/72.
5. A. Hall and R. Fagan, "Definition of System", p.81.
6. W. Buckley, *Sociology and Modern Systems Theory*, p.41.
7. W. Michelson, *Man and His Urban Environment*, pp.62–63.
8. J. Milsum, "The Hierarchical Basis for Modern Living Systems", pp.161–65.

CHAPTER 5—RELATIONSHIP BETWEEN LIFESTYLE AND ETHICAL SUB-SYSTEMS

1. I. Illich, *Alternative to Schooling*, p.32.
2. I. Illich, *Deschooling Society*, p.35.
3. G. Hardin, *Exploring New Ethics for Survival*, p.134.
4. S. Parker, *The Future of Work and Leisure*, p.11.
5. D. Reisman, "Leisure and Work in Post-Industrial Society", p.78.
6. A. Toynbee, *Surviving the Future*, p.94.

7. Ibid., pp.95–96.
8. D. Gabor, *Inventing the Future*, p.120.

CHAPTER 6—RELATIONSHIP BETWEEN LIFESTYLE AND TECHNOLOGICAL SUBSYSTEMS

1. L. White, *Medieval Technology and Social Change*, p.38.
2. L. Sharp, "Steel Axes for Stone-Age Australians", pp.69–90.
3. Ibid., pp.85–86.
4. L. Brown and G. Finsterbusch, "Man's Quest for Food", p.258.
5. M. Allaby, "Green Revolution: Social Boomerang", pp.18-21.
6. P. Ehrlich and A. Ehrlich, *Population, Resources and Environment*, pp.121-22.
7. I. Illich, *Celebration of Awareness*, p.133.
8. M. Mead, Interview, "Monday Conference", A.B.C. television.
9. A. Toffler, *Future Shock*, pp.222-23.
10. M. McLuhan and G. Leonard, "The Future of Sex", pp.56-63.
11. B. Skinner, *Beyond Freedom and Dignity*.
12. L. Yablonsky, *Robopaths*, pp.91-92.
13. C. Reich, *The Greening of America*.
14. T. Roszak, *The Making of a Counterculture*.
15. H. Pauli, "Relationships between Land Use and Water Pollution", pp.565-80.
16. F. Klemm, *A History of Western Technology*, p.30.
17. Ibid., pp.38-39.
18. R. Fuller, "Preview of building".

CHAPTER 7—RELATIONSHIP BETWEEN LIFESTYLE AND ENVIRONMENTAL SUBSYSTEMS

1. A. Rapoport, "Environment and People", p.12.
2. For example, F. Thistlewaite, *The Great Experiment*, pp.59-68.
3. N. Pole, "An Interview with Paul Ehrlich", pp.18-19.
4. R. Dubos, "Man and His Environment", pp.688-89.
5. M. Nicholson, "Man's Use of the Earth", pp.10-21.
6. "Report of the Study of Critical Environmental Problems", *Man's Impact on the Global Environment* (Cambridge, Mass.: M.I.T. Press, 1970), p.22.
7. F. Clarkson et al., "Family Size and Sex Role Stereotypes", pp.254-58.
8. D. Stott, "Cultural and Natural Checks on Population".
9. J. Calhoun, "The Social Aspects of Population Dynamics", pp.139-59.
10. R. Hofstadter, *The Progressive Historians*, pp.157-59.
11. A. Rapoport, *Australia as Human Setting*, p.4.
12. R. Theobald, "Planning with People", p.184.
13. E. Fromm, *The Revolution of Hope*, p.1.
14. C. Reich, *The Greening of America*, pp.161-83.

CHAPTER 8—RELATIONSHIP BETWEEN ETHICAL AND ENVIRONMENTAL SUBSYSTEMS

1. For example, J. Passmore, *Man's Responsibility for Nature*.
2. A. Spilhaus, "Ecolibrium", p.713.

3. B. Brown, "Pollution", p.351.
4. P. Dwyer and J. Harris, "The Ecologist as Conservationist", p.26.
5. C. Birch, "Blueprint for Survival", Symposium on Survival S.O.S. '73, Queensland Conservation Council, Brisbane 1973.
6. C. Stone, *Should Trees Have Standing?*
7. J. Mosley, "Towards a History of Conservation in Australia", p.153.
8. "The Case against Pike Creek Dam", Queensland Conservation Council and the University of Queensland Speleological Society, Brisbane, 1973.
9. J. Swan, "Environmental Education", pp.225-26.
10. J. Fischer, "Survival U", pp.12-22.
11. W. Bruvold, "Belief and Behaviour as Determinants of Environmental Attitudes", pp.209-10.
12. R. Heathcote, "The Visions of Australia", pp.77-98.
13. J. Freeland, in Rapoport, p.108.
14. R. Heathcote, in Rapoport, p.79.
15. *The Nimbin Good Times* (Nimbin: Coordination Co-operative Ltd., 1974).
16. Bruvold, pp.209-10.
17. S. Wapner et al., *Environment and Behaviour*, P.259.

CHAPTER 9—RELATIONSHIP BETWEEN ETHICAL AND TECHNOLOGICAL SUBSYSTEMS

1. J. Ellul, *The Technological Society*.
2. A. Toffler, *Future Shock*.
3. R. Heilbroner, *The Future of History*, chap. 2.
4. H. Marcuse, *One-Dimensional Man*.
5. C. Reich, *The Greening of America*.
6. T. Roszak, *The Making of a Counterculture*.
7. J. Douglas, "Freedom and Tyranny in a Technological Society", p.31.
8. Ibid., p.12.
9. F. Klemm, *A History of Western Technology*, p.17.
10. F. Fraser Darling, "The Technological Exponential", p.170.
11. *Australian*, 21 February 1974, p.14.
12. Matheson, "Letter to the Editor", *Australian*, 1 March 1974, p.4.
13. E. Mesthene, "How Technology Will Shape the Future", p.137.
14. R. Brasch, *How Did Sex Begin?*
15. G. Taylor, *Sex in History*, p.157.
16. For example, C. Westoff and R. Rindfuss, "Sex Preselections in the United States", pp.633-36; M. Shaw, "Genetic Counselling", p.751.
17. A. Etzioni, "Sex Control, Science, and Society", pp.83-89.
18. J. Postgate, "Bat's Chance in Hell", p.15.
19. H. Marcuse, "The New Forms of Control".
20. J. Wilkinson, in Douglas, pp.49-50.

CHAPTER 10—RELATIONSHIP BETWEEN TECHNOLOGICAL AND ENVIRONMENTAL SUBSYSTEMS

1. L. Orleans and R. Suttmeier, "The Mao Ethic", pp.1,174.
2. W. Westman, "Environmental Impact Statements", pp.465-70.
3. "We Are Fifteen", p.20.
4. H. Conklin, *The Study of Shifting Cultivation*.

5. P. Dansereau, "Ecology and the Escalation of Human Impact", p.634.
6. O. Stewart, "Fire as the First Great Force Employed by Man", pp.126-27.
7. M. Nicholson, "Man's Use of the Earth", p.13.
8. E. Anderson, "Man as a Maker of New Plants and New Plant Communities", pp.653-60.
9. T. Jacobsen and R. Adams, "Salt and Silt in Ancient Mesopotamian Agriculture", pp.383-94.
10. R. Bryson and D. Baerreis, "Possibilities of Major Climatic Modification and Their Implications", pp.136-42.
11. H. Pauli, "Relationships between Land Use and Water Pollution".
12. For example, M. Ozonio de Almeida et al., *Environment and Development*.
13. B. Commoner, "Dispute", pp.25, 40-52; *The Closing Circle*.
14. M. Scheidt, "Environmental Effects of Highways", pp.419-27; E. Ullman, "The Role of Transportation and the Bases for Interaction", pp.862-80.
15. A. Wolman, "Disposal of Man's Wastes", pp.807-16.
16. H. Kahn and A. Weiner, *The Year 2,000*.
17. R. Fuller, "An Operating Manual for Spaceship Earth", pp.341-89.
18. I. Rubinoff, "Central American Sea-Level Canal", pp.493-501.
19. "Isis District of Queensland: A Land Use Study", Interdepartmental Committee, Queensland Government, May 1971.
20. S. Weissman and K. Barkley, "The Eco-Establishment", pp.48-49,54,56,58.

CHAPTER 11—SELF-INDUCED CHANGES WITHIN SUBSYSTEMS

1. R. Russell, "Environmental Changes through Forces Independent of Man".
2. J. Gribben and S. Plagemann, *The Jupiter Effect*.
3. D. Tarling and M. Tarling, *Continental Drift*.
4. B. Thom, "Coastal Erosion in Eastern Australia", pp.198-209.
5. E. Graham, "The Recreative Role of Plant Communities", pp.677-91.
6. M. Burnett, *Natural History of Infections and Diseases*, p.308.
7. J. Henningham et al., *Australian*, 19 June 1974, p.7.
8. R. Bryson, "A Perspective on Climatic Change", pp.753-60.
9. *United Nations in Action*, no.74/17 (24 May 1974), p.2.
10. H. Lamb, *The Biological Significance of Climatic Change in Britain*, p.3.
11. G. Taylor, *The Doomsday Book*, pp.54-77; Bryson, 753-60.
12. C. Reich, *The Greening of America*.
13. T. Roszak, *The Making of a Counterculture*.
14. L. Yablonsky, *Robopaths*.
15. For example, R. Bell, *The Sex Survey of Australian Women*.
16. For example, G. Taylor, *Sex in History*; R. Brasch, *How Did Sex Begin?*; G. Taylor, *The Angel Makers*.
17. W. Wertheim, *Evolution and Revolution*, p.17.
18. H. Barringer et al., *Social Change in Developing Areas*, p.21.
19. Wertheim, p.20; B. Moore, *Social Origins of Dictatorship and Democracy*.
20. H. Kahn and B. Bruce-Briggs, *Things to Come*.
21. A. Toffler, *Future Shock*, p.34.
22. L. White, *Medieval Technology and Social Change*, p.63.
23. S. Lilley, *Men, Machines, and History*, pp.308-9.

CHAPTER 12—THE HEURISTIC MODEL: REVIEW AND OBSERVATIONS

1. J. Milsum, "The Heirarchical Basis for General Living Systems", p.162.
2. E. Laszlo, "Introduction", pp.10-11.
3. For example, M. Mesarovic, "A Mathematical Theory of General Systems", pp.252-57.

CHAPTER 13—APPLICATION AND UTILIZATION OF THE HEURISTIC MODEL

1. G. Carefoot and E. Sprott, *Famine on the Wind*, p.14.
2. H. Kahn and B. Bruce-Briggs, *Things to Come*, pp.154,158.
3. C. Clark, *Starvation or Plenty; Population Growth and Land Use.*
4. E. Goldsmith et al., "A Blueprint for Survival", p.4.
5. P. Ehrlich and A. Ehrlich, *Population, Resources, Environment*, p.123.
6. For example, D. Johnson, *World Agriculture in Disarray*; M. Hedgcock, "Ethiopia Casts Off Feudal Curbs", *Australian*, 23 March 1974, p.11; "Folly and Famine", *Nation Review*, 29 March–4 April, 1974, p.751.
7. M. Lipton, "The International Diffusion of Technology", p.48.
8. S. Enke, "Birth Control for Economic Development", p.194.
9. W. Paddock, *Famine 1975!.*
10. Ibid., p.209.
11. For example, G. Conway, "Man-Made Plagues", pp.16-18; D. Pimentel, "Realities of a Pesticide Ban", pp.18-31.

Bibliography

Aarons, E. *Philosophy for an Exploding World*. Sydney: Brolga, 1972.

Adelman, C. *Generations: A Collage on Youthcult*. Ringwood, Vic.: Penguin, 1972.

Agron, C. "Some Observations on Behaviour in Institutional Settings". *Environment and Behaviour* 3, no.1 (March 1971): 103-14.

Allaby, M. "Green Revolution: Social Boomerang". *Ecologist* 1, no.3 (1971): 18-21.

———. "Miracle Rice and Miracle Locusts". *Ecologist* 3 (May 1973): 180-85.

Allsop, B. *The Garden Earth: The Case for Ecological Morality*. New York: Morrow, 1972.

Anderson, K. "Man as a Maker of New Plants and New Plant Communities". In *Man's Impact on Environment*, edited by T.R. Detwyler, pp.653-66. New York: McGraw Hill, 1971.

Artin, T. *Earth Talk*. New York: Grossman, 1973.

Australian Public Opinion Poll 1, no.1 (September 1971).

Banner, D. "Pollution: Symptom of a Value Crisis". *Training and Development Journal* 24, no.12 (December 1970): 16-19.

Barnes, R. "Birth Control in Popular Twentieth Century Periodicals". *The Family Coordinator*, April 1970, pp.159-64.

Barringer, H. et al. *Social Change in Developing Areas*. Cambridge, Mass.: Schenkman, 1965.

Bass, B. "Conformity, Deviation, and a General Theory of Interpersonal Behaviour". In *Conformity and Deviation*, edited by I. Berg and B. Bass, pp.38-100. New York: Harper and Brothers.

Bassett, I. et al. "Changes in Atmospheric Ozone and Solar Ultraviolet". *Search* 5, no.5 (May 1974): 182-86.

Becker, H. *Outsiders: Studies in the Sociology of Deviance*. New York: The Free Press of Glencoe, 1963.

Beeton, A. "Eutrophication of the St. Laurence Great Lakes". In *Man's Impact on Environment*, edited by T. Detwyler, pp.233-45. New York: McGraw Hill, 1971.

Behrman D. Exerpt from *Realities* 259 (1972): 27.

Bell, R. *The Sex Survey of Australian Women*. Melbourne: Sun Books, 1974.

Bernard, H. and Pelto, P. *Technology and Social Change*. New York: Macmillann 1972.

Berry, A. *The Next Ten Thousand Years*. New York: Saturday Review Press/E.P. Dutton, 1974.

Berton, P. *The Smug Minority*. Toronto: McClelland and Stewart, 1968.

Birch, C. 'People Are Dying Because of the Air They Breathe". *Current Affairs Bulletin* 49, no.10 (March 1973): 318-19.

Blake, J. "Population Policy for Americans: Is the Government Being Misled?". *Science* 164 (May 1969): 522-29.

Blake, R. and Mouton, J. "The Experimental Investigation of Interpersonal Choice". In *The Manipulation of Human Behaviour*, edited by A. Biderman and H. Zimmer, pp.216-76. New York: John Wiley and Sons, 1961.

Bloom, H. and Springell, P. "The Lead in Petrol Controversy Comes to Australia". *Search* 4, no.5 (May 1973): 152-54.

Blumberg, P. *Industrial Democracy: The Sociology of Participation*. London: Constable, 1968.

Bookchin M. "Toward an Ecological Solution". *Ramparts* 8, no.11 (May 1970): 7-15.

Boulding, K. "The Economics of the Coming Spaceship Earth". In *Global Ecology*, edited by J. Holdren and P. Ehrlich, pp.180-92. New York: Harcourt Brace, 1971.

Braithwaite, J. Submission to National Population Enquiry. Canberra, 1973.

Braithwaite, J. et al. "A Test of Reich's Typology Using Image Analysis". Mimeograph. Department of Anthropology and Sociology, University of Queensland, 1973.

Brasch, R. *How Did Sex Begin?* Sydney: Angus and Robertson, 1973.

"Brook Farm". In *American Utopianism*, edited by R. Fogarty, pp.62-69. Ithacan Ill.: Peacock Publishers, 1972.

Brower, K. "The American Wilderness". In *The Environmental Handbook*, edited by J. Barr, pp.41-46. London: Ballantine, 1971.

Brown, B. 'Pollution: Entropy in the Social Engine". *Southern Quarterly* 8, no.4 (July 1970): 349-56.

Brown, B. and Finsterbusch, G. "Man's Quest for Food: Its Ecological Implications". In *Politics and Environment*, edited by W. Anderson, pp.253-65. Pacific Palisades: Goodyear, 1970.

Bruvold, W. "Belief and Behaviour as Determinants of Environmental Attitudes". *Environment and Behaviour* 5, no.2 (June 1973): 202-18.

Bryson, R. "A Perspective on Climatic Change". *Science* 184, no.4,138 (17 May 1974): 753-60.

Bryson, R. and Baerreis, D. "Possibilities of Major Climatic Modification and Their Implications: Northwest India, A Case for Study". *Bulletin of American Meteorological Society* 48, no.3 (March 1967): 136-42.

Buckhout, R. "The War on People". *Environment and Behaviour* 3, no.3 (September 1971): 322-44.

Buckley, W. *Sociology and Modern Systems Theory*. Englewood Cliffs, N.J.: Prentice-Hall, 1967.

Buechner, H. "The Ecosystem Level of Organization", In *Organized Systems in Theory and Practice*, edited by P. Weiss, pp.45-58. New York: Hafner, 1971.

Burnett, M. *Natural History of Infections and Diseases*. Cambridge, Mass.: Harvard University Press, 1962.

Calhoun, J. "The Social Aspects of Population Dynamics". *Journal of Mammalogy* 33 (1952): 139-59.

Carefoot, G. and Sprott, E. *Famine on the Wind*. New York: Rand McNally and Co., 1967.

Chapman, D. "An End to Chemical Farming". *Environment* 15, no.2 (March 1973): 12-17.

Chase, S. *The Most Probable World*. Baltimore, Maryland: Penguin, 1968.

Clark, C. *Population Growth and Land Use*. London: Macmillan, 1967.

———. "Population Growth May Lead to Excessive Wealth". *Current Affairs Bulletin* 49, no.15 (1 March 1973): 314-17.

———. *Starvation or Plenty*. London: Secker and Warburg, 1970.

Clark, C. and Llewellen-Jones, D. *Zero Population Growth*. Adelaide: Heinemann, 1974.

Clarkson, F. et al. "Family Size and Sex Role Stereotypes". In *Family in Transition*, edited by A. & J. Skolnick, pp.254-58. Boston: Little Brown, 1971.

Coale, A. "Man and His Environment". *Science* 170, no.3,954 (9 October 1970): 132-36.

Commoner, B. "Dispute". *Environment* 14, no. 3 (April 1972): 25, 40-52.

———. *The Closing Circle*. London: Jonathon Cape, 1972.

Conklin, H. *The Study of Shifting Cultivation*. Washington: Pan American Union, 1963.

Conway, G. "Man-Made Plagues". *Ecologist* 1, no.4 (1970): 16-18.

Coombs, H. "Matching Ecological and Economic Realities". Prepared by the Australian Conservation Foundation, Parkville, Vic., 1972.

Dansereau, P. "Ecology and the Escalation of Human Impact". *International Social Science Journal* 22, no.4 (1970): 628-47.

Department of Social and Preventive Medicine, University of Queensland. A Study of Vasectomy in Brisbane. Project 328/72.

Desmond, A. "How Many People Have Ever Lived on Earth?". In *The Population Crisis*, edited by L. Ng and S. Mudd, pp.20-38. Bloomington: Indiana University Press, 1965.

Detwyler, T. "Summary and Prospect". In *Man's Impact on Environment*, edited by T. Detwyler, pp.695-700. New York: McGraw Hill, 1971.

Douglas, J. "Freedom and Tyranny in a Technological Society". *Freedom and Tyranny*, edited by J. Douglas, pp.3-30. New York: Alfred A. Knopf, 1970.

Dubos, R. "Environmental Biology". *Bioscience* 14, no.1 (January 1964): 11-14.

———. "Man and His Environment: Scope, Impact, and Nature". In *Man's Impact on Environment*, edited by T. Detwyler, pp.684-94. New York: McGraw Hill, 1971.

Dumont, R. and Rosier, B. *The Hungry Future*. New York: Praeger, 1969.

Dwyer, P. and Harris, J. "The Ecologist as Conservationist". *Search* 4, nos.1-2 (January-February 1973): 24-28.

Ecology East Action. "The Power to Destroy, the Power to Create". In *The Ecological Conscience*, edited by R. Disch, pp.161-69. New York: Prentice-Hall, 1970.

Edgar, D., ed. *Social Change in Australia*. Melbourne: Cheshire, 1974.

Ehrlich, P. *The Population Bomb*. London: Ballantine, 1968.

Ehrlich, P. and Ehrlich, A. *Population, Resources, Environment*. San Francisco: W.H. Freeman and Co., 1972.

———. "The Food from the Sea Myth". *Saturday Review* 53 (4 April 1970): 55.

Ehrlich, P. et al. "Dispute". *Environment* 14, no.3 (April 1972): 24-52.

———. *Human Ecology*. San Francisco: W.H. Freeman and Co., 1973.

Ehrlich, P. and Harriman, R. *How to be a Survivor*. London: Ballantine, 1971.

Ehrlich, P. and Holdren, J. "Impact of Population Growth". *Science* 171 (26 March 1971): 1212-16.

Eisner, T. et al. "Population Control, Sterilization, and Ignorance". *Science* 167, no.3,917 (23 January 1970): 337.

Ellul, J. *The Technological Society*. New York: Alfred A. Knopf. 1964.

Enke, S. "Birth Control for Economic Development". *Global Ecology*, edited by J. Holdren and P. Ehrlich, pp.193-99. New York: Harcourt Brace, 1971.

Etzioni, A. "Sex Control, Science, and Society". In *Family in Transition*, edited by A. and J. Skolnick, Boston: Little Brown and Co., 1971, pp.83-89.

"Family Planning in Australia". *Medical Journal of Australia* 2, no.10 (8 September 1973): 473-74.

Figa-Talamanca, I. "Inconsistencies of Attitudes and Behaviour in Family Planning Studies". *Journal of Marriage and the Family* 34, no.2 (May 1974): 336-44.

Fischer, J. "Survival U: Prospects for a Really Relevant University". *Harpers*, September 1969, pp.12-22.

Fogarty, R., ed. *American Utopianism*. Itasca, Ill.: Peacock Publishers, 1972.

Food and Agriculture Organization. *The State of Food and Agriculture*. Rome: F.A.O., 1971.

Fraser Darling, F. "The Technological Exponential". In *The Environmental Handbook*, edited by J. Barr, pp.169-76. London: Ballantine/Friends of the Earth, 1971.

Freeland, J. *Architecture in Australia*. Ringwood, Vic.: Penguin, 1968.

———. "People in Cities". In *Australia as Human Setting*, edited by A. Rapoport, pp.99-123. Sydney: Angus and Robertson, 1972.

Frejka, T. "The Prospects for a Stationary World Population". *Scientific American* 228, no.3 (March 1973): 15-23.

Fromm, E. *The Revolution of Hope*. New York: Harper and Row, 1968.

Fuller, R.B. "An Operating Manual for Spaceship Earth". In *Environment and Change, The Next 50 Years*, edited by W. Ewald, pp.341-89. Bloomington: Indiana University Press, 1968.

———. "Preview of Building". In *The Buckminster Fuller Reader*, edited by J. Mellor, pp. 285-309. Ringwood, Vic.: Penguin, 1972.

Gabor, D. *Inventing the Future*. Ringwood, Vic.: Penguin, 1963.

Garbutt, H. "Mental Health Services: Agent of Social Control?". *Iconoclast* 5 (1972): 5-10; and 6 (1973): 5-12.

Goldman, J. "Toward a National Technology Policy". *Science* 177 (22 September 1972): 1,078-80.

Goldman, M. "The Convergence of Environmental Disruption". *Science* 170, no.3, 953 (2 October 1970): 37-42.

Goldsmith, E. et al. "A Blueprint for Survival". *Ecologist* 2, no.1 (1972).

Graham, E. "The Recreative Role of Plant Communities". In *Man's Role in Changing the Face of the Earth*, edited by W. Thomas et al., pp.677-91. Chicago: University of Chicago Press, 1956.

Gribben, J. and Plagemann, S. *The Jupiter Effect*, New York: Walker and Co., 1974.

Gross, A. "Vasectomy". *Forum* 1, no.3 (1973): 10-12.

Hagevik, G. and Mann, L. "The 'New' Environmentalism: An Intellectual Frontier". *American Institute of Planners Journal* 37, no.4 (July 1971): 274-80.

Hall, A. and Fagen, R. "Definition of System". *Modern Systems Research for the Behavioural Scientist*, edited by W. Buckley, pp.81-92. Chicago: Aldine Publishing Co., 1968.

Hamilton, D. *Technology, Man, and the Environment*. London: Faber and Faber, 1973.

Hardin, G. *Exploring New Ethics for Survival*. Baltimore: Penguin, 1973.

———. "The Tragedy of the Commons". *Science* 162 (11 December 1968): 1,243-48.

Harkavay, O. et al. "Family Planning and Public Policy: Who Is Misleading Whom?". *Science* 165 (July 1969): 367-73.

Harris, M. *The Rise of Anthropological Theory*. London: Routledge and Kegan Paul, 1969.

Harris, T. "All the World's a Box". *Psychology Today*. (August 1971): 33-35.

Heathcote, R. "The Visions of Australia, 1770-1970". In *Australia as Human Setting*, edited by A. Rapoport, pp.77-98. Sydney: Angus and Robertson, 1972.

Hedgepeth, H. and Stock, D. *The Alternative: Communal Life in New America*. New York: Macmillan, 1970.

Heilbroner, R. *The Future of History*. New York: Harper Torchbooks, 1968.

Henkin, H. "Side Effects". *Environment* 11, no.1 (January-February 1969): 1-8.

Hickey, R. "Air Pollution". *Environment: Resources, Pollution and Society*, edited by W. Murdock, pp.189-212. Stanford, Conn.: Sinaur Associates, 1971.

Hofstadter, R. *The Progressive Historians*. New York: Knopf, 1968.

Horowitz, I. "The Environmental Cleavage: Social Ecology Versus Political Economy". *Social Theory and Practice* 2, no.1 (1972): 125-34.

Hutchinson, J. "The Green Revolution: Hope or Disaster". In *Environmental Solutions*, edited by N. Pole, pp.119-21. Cambridge: Eco Publication, 1972.

Illich, I. *Alternatives to Schooling*. Melbourne: Australian Union of Students, 1972.

———. *Celebration of Awareness*. Ringwood, Vic: Penguin, 1970.

———. *Deschooling Society*. Ringwood, Vic.: Penguin, 1971.

Jacobsen, T. and Adams, R. "Salt and Silt in Ancient Mesopotamian Agriculture". In *Man's Impact on Environment*, edited by T. Detwyler, pp. 383-94. New York: McGraw Hill, 1971.

Johnson, D. *World Agriculture in Disarray*. London: Macmillan, 1973.

Kahn, H. and Bruce-Briggs, B. *Things to Come*. New York: Macmillan, 1972.

Kahn, H. and Wiener, A. *The Year 2000*. New York: Macmillan, 1967.

Kates, R. "Human Perception of the Environment". *International Social Science Journal* 5, 22, no.4 (1970): 648-60.

Keyfitz, N. "The Numbers and Distribution of Mankind". *Environment: Resources, Pollution and Society*, edited by W. Murdock, pp.31-52. Stanford, Conn.: Sinaur Associates, 1971.

Klemm, F. *A History of Western Technology*. Cambridge: MIT Press, 1964.

Kroeber, A. "Relations of Environmental and Cultural Factors". In *Environment and Cultural Behaviour*, edited by A. Vayda, pp.350-60. New York: Natural History Press, 1969.

Lamb, H. *The Biological Significance of Climatic Changes in Britain*. London: Academic Press, 1965.

Laszlo, E. "Introduction". In *The Relevance of General Systems Theory*, edited by E. Laszlo, pp. 1-11. New York: George Braziller, 1972.

Lehrman, N. "Playboy Interview: Germaine Greer". *Playboy* 19, no.1 (January 1972): 61-82.

Leiss, W. "Utopia and Technology: Reflections on the Conquest of Nature". *International Social Science Journal* 22, no.4 (1970): 576-88.

Leopold, A. *A Sand County Almanac*. London: Oxford University Press, 1970.

Lilley, S. *Men, Machines, and History*. London: Laurence and Wishart, 1965.

Lindsay, N. *Bohemians of the Bulletin*. Sydney: Angus and Robertson, 1965.

Lipton, M. "The International Diffusion of Technology". In *Development in a Divided World*, edited by D. Seers and L. Joy, pp.45-63. Ringwood, Vic.: Penguin, 1971.

Loraine, J. *Sex and the Population Crisis*. London: Heinemann Medical Books, 1970.

Luce, G. *Body Time*. Melbourne: Sun Books, 1972.

Lyness, J. et al. "Living Together: An Alternative to Marriage". *Journal of Marriage and the Family* 34, no.2 (May 1972): 305-11.

Mackay, D. "The Informational Analysis of Questions and Commands". In *Information Theory: Fourth London Symposium*, edited by Colin Cherry, pp.469-76. London: Butterworth, 1961.

Maddock, K. *The Australian Aborigines*. Ringwood, Vic.: Penguin, 1974.

Maddox, J. "A Great Many of the More Scary Arguments of Ecological Catastrophe Have Been Shown to Be Chimerical". *Current Affairs Bulletin* 49, no.10 (1 March 1973): 311.

———. *The Doomsday Syndrome*. London: Macmillan, 1972.

Malthus, T. *First Essay on Population 1798*. New York: A.M. Kelley, 1965.

Man's Impact on the Global Environment: Report of the Study of Critical Environmental Problems. Cambridge, Mass.: MIT Press, 1970.

Marcuse, H. *One-Dimensional Man: The Ideology of Industrial Society*. London: Sphere Books, 1968.

——. "The New Forms of Control". In *Freedom and Tyranny*, edited by J. Douglas, pp.33-47. New York: Knopf, 1970.

Marx, L. "American Institutions and Ecological Ideals". *Science* 170, no.3,961 (27 November 1970): 945-52.

McEvoy, J. "The American Concern with the Environment". In *Social Behaviour, Natural Resources and the Environment*, edited by W. Burch, pp.214-36. New York: Harper and Row, 1972.

McLuhan, M. *Understanding Media*. Toronto: Signet, 1964.

McLuhan, M. and Leonard, G. "The Future of Sex". *Look* 31 (25 July 1967): 56-63.

McMichael, T. "The Eco-Crisis in Perspective". *Dissent* 26 (Summer 1971): 3-7.

Meadows, D. et al. *The Limits to Growth*. London: Earth Island, 1972.

Means, R. "Insights into Pollution". *American Institute of Planners*, July 1971, pp.211-17.

——. "Public Opinion and Planned Changes in Social Behaviour: The Ecological Crisis". In *Social Behaviour, Natural Resources and the Environment*, edited by W. Burch, pp.203-13. New York: Harper and Row, 1972.

Meeker, J. "The Comedy of Survival". *Ecologist* 3, no.6 (1973): 210-15.

Meier, R. "Insights into Pollution". *Journal of American Institute of Planners*, July 1971, pp.211-17.

Mellor, J., ed. *The Buckminster Fuller Reader*. Ringwood, Vic.: Penguin, 1970.

Mesarovic, M. "A Mathematical Theory of General Systems". In *Trends in General Systems Theory*, edited by G. Klir, pp.251-69. New York: John Wiley and Sons, 1972.

Mesthene, E. "How Technology Will Shape the Future". In *Environment and Change*, edited by W. Ewald, pp.132-52. Bloomington: Indiana University Press, 1968.

Metcalf, W. "Ethics and Criminal Law". *Iconoclast* 7 (1974): 4-9.

——. "The Counter-Culture—and Ecological Morality". *Trephine* 73 (1973): 41-44.

——. "The Environment: Nixon's New Issue". *Eco Info* 1, no.5 (December 1973): 4-7.

——. "The Ethic of Frugality". *Eco Info* 1, no.3 (June-July 1973): 30-32.

Michelson, W. *Man and His Urban Environment: A Sociological Approach*. Reading, Mass.: Addison Wesley Publishing Co., 1970.

Micklin, P. "Soviet Plans to Reverse the Flow of Rivers". In *Man's Impact on Environment*, edited by T. Detwyler, pp.302-18. New York: McGraw Hill, 1971.

Milsum, J. "The Hierarchical Basis for General Living Systems". In *Trends in General Systems Theory*, edited by G. Klir, pp.145-87. New York: John Wiley and Sons, 1972.

Moncrief, L. "The Cultural Basis for Our Environmental Crisis". *Science* 170, no.3,957 (30 October 1970): 508-12.

Moore, B. *Social Origins of Dictatorship and Democracy*. London: Allen Lane, The Penguin Press, 1966.

Morrison, D. et al. "The Environmental Movement: Some Preliminary Observations and Predictions". In *Social Behaviour, Natural Resources and the Environment*, edited by W. Burch, pp.259-79. New York: Harper and Row, 1972.

Mosley, J. "Towards a History of Conservation in Australia". In *Australia as Human Setting*, edited by A. Rapoport, pp. 136-54. Sydney: Angus and Robertson, 1972.

Musil, J. "Social Change and Environment". *International Social Science Journal* 22, no.4 (1970): 589-606.

Nicholson, M. "Man's Use of the Earth: Historical Background". In *Man's Impact on Environment*, edited by T. Detwyler, pp. 10-21. New York: McGraw Hill, 1971.

Oglesby, C. "The Young Rebels". In *Environment and Change, the Next 50 Years*, edited by W. Ewald, pp. 153-66. Bloomington: Indiana University Press, 1968.

Orleans, L. and Suttmier, R. "The Mao Ethic and Environmental Quality". *Science* 170, no.3,963 (11 December 1970): 1173-76.

Osborn, F. *Our Crowded Planet*. London: Allen and Unwin, 1963.

Ozonio de Almeida, M. et al. *Environment and Development*. New York: Carnegie Endowment for International Peace, 1972.

Paddock, W. "How Green is the Green Revolution?". Bioscience 20, no.16 (15 August 1970): 897-902.

Paddock, W. and Paddock, P. *Famine 1975!*. London: Weidenfeld and Nicolson, 1967.

Parker, S. *The Future of Work and Leisure*. London: Paladin, 1972.

Passmore, J. *Man's Responsibility for Nature*. London: Duckworth, 1974.

Pauli, H. "Relationships between Land Use and Water Pollution". Symposium on Water Pollution, University of Queensland, 1972, pp.5.1-5.28.

Pimentel, D. "Realities of a Pesticide Ban". *Environment* 15, no.2 (1973): 18-31.

Platt, J. "What We Must Do". In *Global Ecology*, edited by J. Holdren and P. Ehrlich, pp.280-91. New York: Harcourt Brace, 1971.

Pohlmon, E. "Population Education: Critical Questions". *Environment and Behaviour* 3, no.3 (September 1971): 310-13.

Pole, N. "An Interview with Paul Ehrlich". *Ecologist* 3, no.1 (January 1973): 18-23.

Postgate, J. "Bat's Chance in Hell". *New Scientist*. (5 April, 1973): 12-15.

President's Science Advisory Panel on the World Food Supply, *The World Food Problem*. Washington: Government Printing Office, 1967.

Price, E. and King, J. "Domestication and Adaptation". In *Man's Impact on Environment*, edited by T.R. Detwyler, pp.640-52. New York: McGraw Hill, 1971.

Queensland Conservation Council and the University of Queensland Speleological Society. *The Case Against Pike Creek Dam*. Brisbane: Queensland Conservation Council and the University of Queensland Speleological Society, 1973.

Queensland Government. Isis District of Queensland: A Land Use Study. Interdepartmental Committee, Queensland Government, May 1971.

Rapid Population Growth: Consequences and Policy Implications. Baltimore: John Hopkins Press, 1971.

Rapoport, A. "Environment and People". In *Australia as Human Setting*, edited by A. Rapoport, pp.3-21. Sydney: Angus and Robertson, 1972.

Reich, C. *The Greening of America*. Ringwood, Vic.: Penguin, 1970.

Reisman, D. "Leisure and Work in Post-Industrial Society". In *The Technological Threat*, edited by Jack Douglas, pp.71-91. Englewood Cliffs, N.J.: Prentice-Hall, 1971.

Rhodes, R. "Sex and Sin in Sheboygan". *Playboy* 19, no.8 (August 1972): 186-90.

Richter, P., ed. *Utopias: Social Ideals and Communal Experiments*. Boston: Holbrook Press, 1971.

Risebrough, R. "Polychlorinated Byphenyls in the Global Ecosystem". *Nature* 220 (1968): 1,098-102.

Roberts, R. *The New Communes: Coming Together in America*. Englewood Cliffs, N.J.: Prentice-Hall, 1971.

Rollin, B. "Motherhood: Who Needs It?" In *Family in Transition*. edited by A. and J. Skolnick, pp.346-56. Boston: Little Brown, 1971.

Rose, A. "Australia as Cultural Landscape". In *Australia as Human Setting*, edited by A. Rapoport, pp.58-74. Sydney: Angus and Robertson, 1972.

Roszak, T. *The Making of A Counterculture*. London: Faber, 1968.

Rubinoff, I. "Central American Sea-Level Canal: Possible Biological Effects". In *Man's Impact on Environment*, edited by T. Detwyler, pp.493-501. New York: McGraw Hill, 1971.

Russell, R. "Environmental Changes through Forces Independent of Man". In *Man's Role in Changing the Face of the Earth*, edited by W. Thomas et al., pp.453-70. Chicago: University of Chicago Press, 1956.

Ryder, W. "Agriculture: The Roots of Deterioration". *New Scientist* 54, no. 799 (8 June 1972): 567-68.

Sauer, C. "The Agency of Man on the Earth". In *Man's Role in Changing the Face of the Earth*, edited by W. Thomas, pp.49-69. Chicago: University of Chicago Press, 1956.

Saunders, D. "Classification of Areas and Districts". Paper Delivered at Seminar of the National Trust of Queensland, 8 December 1973.

Scheidt, M. "Environmental Effects of Highways". In Man's Impact on Environment, edited by T. Detwyler, pp.419-27. New York: McGraw Hill, 1971.

Schell, O. "China's Way with Wastes". *Ecologist* 3, no.2 (February 1973): 56-59.

Sears, P. "Utopias and the Living Landscape". In *Utopias and Utopian Thought*, edited by Frank Manuel, pp.134-42. Boston: Beacon Press, 1967.

Sewell, W. "Environmental Perceptions and Attitudes of Engineers and Public Health Officials". *Environment and Behaviour* 3, no.1 (March 1971): 34-58.

Sewell, W. and Foster, N. "Environmental Revival". *Environment and Behaviour* 3, no.2 (June 1971): 123-34.

Sharp, L. "Steel Axes for Stone-Age Australians". In *Human Problems in Technological Change*, edited by E. Spicer, pp.69-90. New York: Russell Sage Foundation, 1952.

Shaw, M. "Genetic Counselling", *Science* 184, no.4,138 (17 May 1974): 751.

Simon, J. "Science Does Not Show That There Is Overpopulation". Mimeographed. University of Illinois, Department of Economics.

──────. "The Effect of Population Growth on Industrialized Economies: A Different Sort of Simulation". Mimeographed. University of Illinois, College of Commerce and Business Administration.

Siu, R. "The Role of Technology in Creating the Environment Fifty Years Hence". In *Environment and Change, the Next 50 years*, edited by W. Ewald, pp.81-98. Bloomington: Indiana University Press, 1968.

Skinner, B. *Beyond Freedom and Dignity*. New York: Bantam, 1971.

──────. "Freedom, Control and Utopia". *Utopias: Social Ideals and Communal Experiments*, edited by P.E. Richter, pp.287-304. Boston: Holbrook Press, 1971.

──────. *Walden Two*. New York: Macmillan, 1948.

Skolimowski, H. "Technology v. Nature". *Ecologist* 3, no.2 (February 1973): 50-55.

Smith, K. *The Malthusian Controversy*. London: Routledge and Kegan Paul, 1951.

Spilhaus, A. "Ecolibrium". *Science* 175, no.4,023 (18 February 1972): 711-15.

Stewart, C. "The Hazards of Population Forecasts". *New Society* 18, no.469 (23 September 1971): 559-62.

Stewart, O. "Fire as the First Great Force Employed by Man". In *Man's Role in Changing the Face of the Earth*, edited by W. Thomas, pp.115-33. Chicago: University of Chicago Press, 1956.

Stokols, D. et al. "Physical, Social, and Personal Determinants of the Perception of Crowding". *Environment and Behaviour* 5, no.1 (March 1973): 87-115.

Stone, C. *Should Trees Have Standing?* Los Altos: Wettion Kaufmann, 1974.

Stott, D. "Cultural and Natural Checks on Population Growth". In *Environment and Cultural Behaviour*, edited by A. Vayda, pp.90-120. New York: Natural History Press, 1969.

Sumner, W. *Folkways*. New York: Ginn and Co., 1906.

Swan, J. "Environmental Education". *Environment and Behaviour* 3, no.3 (September 1971): 223-30.

Swift, L. *Brook Farm*. New York: Corinth Books, 1961.

Talbot, R. and Dickson, T. "Irrigation Quality of Some Stream Waters in the Lockyer Valley, South Eastern Queensland". *Queensland Journal of Agricultural and Animal Sciences 26* (1969): 565-80.

Tarling, D. and Tarling, M. *Continental Drift*. Ringwood, Vic.: Penguin, 1971.

Taylor, G. *Sex in History*. London: Panther, 1953.

——. *The Angel Makers*. London: Heinemann, 1958.

——. *The Doomsday Book*. London: Panther, 1970.

Theobald, R. "Planning with People". In *Environment and Change*, edited by W. Ewald, pp.182-85. Bloomington: Indiana University Press, 1968.

Thistlewaite, F. *The Great Experiment*. London: Cambridge University Press, 1955.

Thom, B. "Coastal Erosion in Eastern Australia". *Search* 5, no.5 (May 1974): 198-209.

Thomis, M. *The Luddites*. Newton Abbot, Devon: David and Charles, 1970.

Thurston, R. "The Moral Capitulation of the Prisoner". *Iconoclast Behind Bars*. School of Sociology, University of New South Wales, Sydney (July 1971): 12-17.

Tinker, J. "Acid Rain: Is Britain Really the Culprit?" *New Scientist* 58, no.844 (1973): 259.

Toffler, A. *Future Shock*. London: Pan, 1970.

Toynbee, A. *Surviving the Future*. London: Oxford University Press, 1971.

Ullman, E. "The Role of Transportation and the Bases for Interaction". In *Man's Role in Changing the Face of the Earth*, edited by W. Thomas, pp.862-80. Chicago: University of Chicago Press, 1956.

United Nations, *United Nations in Action* 74/17 (24 May 1974), p.2.

Wapner, S. et al. "An Organismic Developmental Perspective for Understanding Transactions of Men and Environment". *Environment and Behaviour* 5, no.3 (September 1973): 255-89.

Wattenberg, B. "Overpopulation as a Crisis Issue: The Nonsense Explosion". In *Pollution Papers*, edited by G. Frankes and C. Solberg, pp.131-43. New York: Meredith Corporation, 1971.

"We Are Fifteen". *Environment* 16, no.1 (January/February 1974): 19, 20, 26-28.

Weisberg, B. "The Politics of Ecology". In *The Ecological Conscience*, edited by Robert Disch, pp.154-60. New Jersey: Prentice-Hall, 1970.

Weiss, P. "The Basic Concept of Hierarchic Systems", In *Hierarchically Organized Systems in Theory and Practice*, edited by P. Weiss, pp.1-43. New York: Hafner, 1971.

Weissman, S. "Why the Population Bomb is a Rockefeller Baby". *Ramparts*, May 1970, pp.42-47.

Weissman, S. and Barkley, K. "The Eco-Establishment". *Ramparts* May 1970: 48-49, 54, 56-58.

Wertheim, W. *Evolution and Revolution*. Ringwood, Vic.: Penguin, 1974.

Westman, W. "Environmental Impact Statements: Boon or Burden". *Search* 4, nos.11-12 (November/December 1973): 465-70.

——. "The Superphosphate Bounty Dispute in Global Perspective". *Eco Info* 2, no.2 (1974): 32-33.

Westman, W. and Clifford, R. "Environmental Impact: Controlling the Overall Level". *Science* 181, no.4,102 (31 August 1973): 819-25.

Westoff, C. and Rindfuss, R. "Sex Preselection in the United States: Some Implications". *Science* 184, no.4,137 (10 May 1974): 633-36.

White, L. *Medieval Technology and Social Change*. Oxford: Oxford University Press, 1962.

——. "The Historical Roots of Our Ecological Crisis". *Science* 155, no.3,767 (10 March 1967): 1,203-07.

White, M. "Computers in the Future", *Search* 4, no.7 (July 1973): 240-43.

Wilkinson, J. "Jacques Ellul as the Philosopher of the Technological Society". In *Freedom and Tyranny*, edited by J. Douglas, pp.48-59. New York: Knopf, 1970.

Wilson, P., ed. *Australian Social Issues of the 70s*. Sydney: Butterworths, 1972.

Wolman, A. "Disposal of Man's Wastes". In *Man's Role in Changing the Face of the Earth*, edited by W. Thomas, pp.807-16. Chicago: University of Chicago Press, 1956.

Wright McKinney, J. "The Individual in a New Industrial Age". *Australian Journal of Social Issues* 8, no.1 (1973): 3-12.

Writing Group of the Tientsin Municipal Revolutionary Committee. "The Industrial Recycling in Red China". In *Environmental Solutions*, edited by N. Pole, pp.122-27. Cambridge: Eco Publications, 1972.

Yablonsky, L. *Robopaths*. Baltimore: Penguin, 1972.

Yancey, W. "Architecture, Interaction, and Social Control". *Environment and Behaviour* 3, no.1 (March 1971): 3-21.

Index

Aarons, Eric, 26
Aborigines: environmental perceptions of, 62; Yir Yorant, 50
Abortion, technology of, 35, 83
Agricultural technology, 90, 91, 92, 119, 125–26; changes in, 120–23; extension problems in, 121, 127; reluctance to change, 119
Agriculture and food supply and the environment, 13, 17–18, 34, 118–23, 125–26
Air pollution: simplistic solutions to, 6; dangers of, 14
Alienation from nature, 66
Application of the model, 117–28
Aquaculture, 14, 93, 122, 126
Aswan Dam, 91
Australia: agricultural technology in, 92; migration to, 61–62, 74
Australian Conservation Foundation, 72
Australian environment: changes in, 90; perceptions of, 61–62, 74–75
Australian Mining Industry Council, 4

Bacon, Francis, 25, 70
Beach erosion, 98
Behavioural modification, 20–21, 53
Benefit/cost study: on birth control, 123–24; on Pike Creek Dam, 72
Berry, A., 22
Birch, Charles, 25–27, 71

Birth control technology: changes in lifestyle from, 53, 101; economic impact of, 123–24; ethical effects of, 39, 83–84, 126, 127; historical development of, 83; imposed, 93, 123; overpopulation and, 34–35, 123, 126, 127; sex predetermination and, 84
Blight in United States corn belt, 121–22
Blueprint for Survival, 121
Boerma, Dr., 18
Bookchin, Murray, 28–29
Borlaug, Norman, 121
Bruvold, W., 77
Bryson, 99
Buckley, Walter, 35
Buechner, H., 34
Bulletin, 75

Carson, Rachael, 3, 76
China: environmental problems of, 26; medical technology in, 88
Christianity: cultural change and, 101–2; effects of, on environment, 25; cultural change and, 101–2; technology and, 6, 81
Clark, Colin, 16, 23, 121
Climate, control of, 91
Climatic changes in environment, 99
Climatologists, predictions of, 99
Cold war, 82
Commoner, Barry, 3, 12, 19, 22–23, 76, 78, 90–91
Communal life, 28; technological effects on, 52. See also Nimbin
Computers: environmental change and, 90; technological change

from, 105
Conservationists: arguments used
by, 72; beliefs of, 76; starvation
and, 119
Coombs, H. C., 24
Counterculture, 54, 100–101; accept-
able technology of, 57. *See also*
Communal living, Nimbin
Cult of the cottage, 65
Cultural adaptability to environment,
63–64
Cultural evolution, 100, 102–3,
106, 113
Culture as lifestyle and ethics, 42,
100

Darling, Fraser, 81
Darwin, Charles, 70, 102
DDT, 9, 14, 88, 90
Depopulation, fears of, 9
Designed environments, 66
Dialectic, concept of, 26
Domestication of plants and animals,
89
Douglas, J., 79
Dubos, René, 3, 20, 63
Dumaresq valley, 72–73
Dynamic equilibrium of the system:
breakdown of, 67–68, 105–7;
tendencies to, 38, 46–47, 59,
66–68, 77, 86, 94–95, 96,
111–16

Earth garden philosophy, 25
Ecological: morality, 77; point of
view, 60; vision, 76
Ecology, first lesson of, 7
Economic man as a heuristic model,
36
Ecosystem, world: definition of, 34;
starvation and, 119; technological
threats to, 94
Education, ethical concepts of, 44
Education for leisure, 46
Ehrlich, Paul, 3, 9, 12, 19, 23, 27–
28, 57, 76
Ellul, Jacques, 78, 85
Emerson, R. W., 25
Engel's law, 85
Enke, S., 124
Entropy, 15
Environment: definition of, 37;
effects of population on. *See*

Population, environmental effects
of; ethical determination of,
70–74; lifestyle determination
of, 64–66; man created, 70–71.
See also Environmental changes,
man made; impact on, 64–66,
69, 88; man's perception of,
24–27, 70–72, 74–77, 96, 119,
127; self-induced changes in,
96–100; technological determination of, 88–91
Environmental action groups, 4, 72,
76
Environmental changes, climatic,
99; evolutionary, 98–99; geological, 98; historical, 89; man
made, 88–91; naturally occuring,
97
Environmental crisis, idea of, 3–8,
118, 119; public opinion polls
and, 3–4
Environmental determinism, 60–61
Environmental education, proposals
for, 73–74
Environmental impact statements,
89, 93–94
Environmental possibilists, 61
Environmental probabilism, 61
Environmental problems, interrelated nature of, 33–34, 122,
126
Environmental protection industry,
5, 92–93
Environmental restraints on technology, 92
Environmental solutions, ideas for,
113–15, 126; reasons for
different types, 29–30
Erie, Lake, 13
Ethical beliefs, dysfunctional, 45
Ethics: derivation of, 44; definition
of, 37; environmental determination of, 74–77; lifestyle determination of, 44–46; mass starvation and, 119; self-induced
changes in, 100–103; technological determination of, 82–85
Etzioni, A., 84
Eutrophication, 88
Evolution of culture. *See* Cutlure
evolution
Evolutionary changes in environment, 98–99, 113

Experimental Farm Cottage, 75

Family size: effects of woman's self concept on, 65, 126; ideas of ideal, 62–63
Feminine stereotypes, effects on family size, 65
Fire as an environmental influence, 89
Fiske, Edward, 25
Food: consumption, patterns of, 123; losses due to disease, 120; supply, allocation of, 124–25; supply, increasing, 120–23
Foreign aid and birth control, 123–25
Freeland, J. M., 75
Freeways, environmental impact of, 65
Freud, S., 36
Fromm, Eric, 67
Frugality, ethic of, 26, 114, 127
Fuller, R. B., 22, 53, 57, 70, 91
Futurology, 45, 53, 91, 94, 97, 103, 121

Gabor, D., 46
Galilei, Galileo, 25
Gardens as created environments, 71
Genetic architecture, 91
Geodesic domes, 57
Geological changes in environment, 98
Gifford, R., 24
Gold, 34
Greek views of technology, 80, 84
Green revolution: cultural effects of, 50, 122; definition of, 50, 126; food supply and, 121–23; India and, 51; problems of, 51, 121–23, 125–26
Greenhouse effect, 11, 94, 100, 114

Heathcote, R., 74–76
Heilbroner, R., 79
Heuristic model: application of, 125–28; definition of, 31, 36; purpose of, 117; testing of, 117; understanding of, 42, 48–49, 59, 69, 79, 87, 96, 111–16
Horowitz, Irving, L., 20
Horse collar, importance of, 49–50, 104
Housing: domes as, 57; environ-

mental impact of, 64–65; styles of, 44, 56–57; technology of, 20, 56–57
Human ecology, 118

Ice age effect, 14, 94, 100, 114
Illich, Ivan, 44, 51
Irrigation: development, 92; effects on crop production, 55, 90, 120–21; effects on soil, 55, 88, 89, 121

Judeo Christianity. See Christianity

Kahn, Hermann, 91, 103, 121
Kwashiorkor disease, 17

Lamb, H., 99
Legal rights for natural objects, 71
Leisure time: problem of. See Work and leisure, concepts of; technology of, 58
Lifestyle: definition of, 37; environment and, 27–29; environmental determination of, 60–64; ethical determination of, 42–44; self-imposed changes in, 100–103; technological determination of, 49–54
Linear (non systems) reasoning, 7, 33, 122–25, 128
Lipton, M., 123
Luddites, 85

McKinney, Judith Wright, 26
McLuhan, Marshall, 52–53, 58
Maddox, John, 16, 23
Malnutrition. See Starvation
Malthus, Thomas, 9
Mao Tse Tung, 26
Marcuse, H., 79, 85
Marriage: changes in, 28; ethical concepts of, 27, 28, 43; serial, 52. See also Nuclear family
Marxism, 102
Mead, Margaret, 52
Meadows, Dennis M., 3
Medical resources, allocation of, 124–25
Michelson, W., 37
Migration movements, 61, 74–75
Minoan civilization, 56, 80
Mobility, environmental effects of, 66

Mosley, J. G., 72
Mundey, Jack, 5

National Organization of None
    Parents (N.O.N.), 62
Natural law, concepts of, 102
Natural resource unit, 23–24
Natural system, 7, 31, 115
Newton, Isaac, 25
Nimbin, 25, 29, 57
Nixon, Richard M., 5
Nuclear family, 27, 43; development
    of, 52; technological effects on,
    52

Oceans, changes in level of, 99
Optimization of system and sub-
    systems, 112, 115
Orleans, Leo, 26
Osborne, Fairfield, 12
Ozone, threats to, 14, 88, 94

Paddock brothers, 124–25
Panama Canal, proposed changes to,
    91
Pathogenic organisms, evolution of,
    98–99
Pesticides, effect on crop production,
    55, 127
Phillips, G. P., 4
Pike Creek Dam, 72–73
Plague, 98–99
Planetary engineering, 91
Planning strategies, framework for,
    114
Plant communities: climax condition
    of, 98; simplification of, 121
Plato, 80, 102
Platt, John, 20
Politics, environmental effects on
    early Australian, 75
Pollution: agricultural, 17, 119, 121,
    126; attitudes toward, 13, 71;
    Chinese, 88; problem of, 13–14
Population: environmental effects
    of, 12–13; perceptions of, 62
Population growth: cultural limits on,
    65–66; Green Revolution and,
    121, 123; predictions of, 9–12;
    problems of, 8–13, 33; re-
    ducing, 123–25
Postgate, J., 84
Preservationist attitude to environ-
    ment, 71

Protein: from the sea, 14, 93, 122,
    126; shortages, 17

Queensland Conservation Council, 4

Racism, 102
Radical ecology, 71, 76
Rapoport, A., 60
Reduction of population, 123–25
Reductionist philosophy in educa-
    tion, 69
Reich, Charles, 26, 54, 67, 79, 100
Reith Lecture, 1969, 81
Religious respect of technology, 85
Research, attitudes toward, 84
Resources: agricultural, 121; amount
    of, 14–16, 33; rate of use of,
    15–16; renewable and non-
    renewable, 14–16
Revolution, social, 103. See also
    Cultural evolution
Robopathology, 53
Roman views of technology, 80–81
Roszak, T., 54, 79, 100
Rousseau, J. G., 25. 70

Sears, Paul, 76
Sex: and technology, 52. See also
    Abortion, technology of; Birth
    control technology; control,
    technology of, 84
Sexual behaviour, changes in, 28,
    100, 101
Sharp, L., 50
Simulation model: purposes of, 36,
    117; testing of, 117
Siu, R. G., 21–22
Skinner, B. F., 20–21, 53
Skolimowski, Henry, 25–26
Social change, antimaterialist bias
    of, 48
Social darwinism, 102
Social evolution. See cultural evolu-
    tion
Solar energy, 15, 29, 93, 126
Solutions to environmental crisis:
    economic, 23–24; philosophical,
    24–27; scientific, 19–21; social
    change, 27–29; technological,
    21–23
Space research, benefits of, 104–5
Speleologists, attitudes of, 72–73
Spilhaus, Athelstan, 21
Starvation: environmental problem

of, 118–19; population control policy of, 124–25; predictions of, 17, 99, 118, 121, 125. *See also* Agricultural technology; problem of mass, 18, 109, 117–18, 127–28; reducing, 119–28
Steady state economy, need for, 45
Stone axes, replacement of, 50
Stott, D. H., 65
Sugar cane: growing, 92; harvesting, 51
Supernatural phenomena and ethics, 101–2
Superphosphate, 57, 121
Supersonic transport, 14, 94
Survival University, 73–74
Suttmier, Richard, 26
Synthetic foods, 122–23
System stability and instability, 38, 112–16, 125–28; general theory of, 113
System survival, goal of, 115
Systems model: application of, 117–28; definition of, 35–37; development of, 36–41; equilibrium and optimization within, 38, 112, 115; goal-oriented nature of, 115; how to change, 113–15; information flows within, 37–39, 116; operation of, 38–40; stability of, 40, 112–16

Taylor, Gordon R., 3
Technocratic society, 81
Technological changes: ancient examples of, 56; European colonization and, 56; housing and, 56; industrial revolution and, 50; medieval, 49–50, 104; time lags in, 55
Technological innovation, process of, 104
Technological subsystem, instability of, 113, 126
Technology: agricultural, changes in, 120–23, changes in, 22, 90–91; communications, 58; definition of, 37; dynamic nature of, 103–4; environmental determination of, 92–93; ethical determination of, 80–82; historical development of, 22, 80–81, 104; lifestyle determination of, 54–58; other influences on, 54–55; perceptions

of, 78–79, 81, 82, 84–85; problems of introduction of, 51; radical critique of, 85; self-induced changes in, 103–6; social issue of, 48; transport, 91; waste disposal, 91; Western, in third world, 90
Texas caves, 72–73
Thalidomide, 55
Thoreau, Henry, 25, 70
Thring, Meredith, 73
Time factors in environmental crisis, 18–19
Toffler, Alvin, 52, 56, 78, 104, 105
Toynbee, Arthur, 45–46
Transcendentalists, 25
Transistors, development of, 105
Transport: development of land, 104; food, 120
Triage, 124–25
Tuntable Falls commune. *See* Nimbin

United Nations predictions of population, 9–12
United States Conservation Foundation, 12
United States of America: Department of Agriculture, 125; President's Advisory Panel on the World Food Supply, 17; President's Population Commission, 63; research allocation, 80; Supreme Court, 71
Urban sprawl, 65
Utilitarian (or transformational) attitude to environment, 71
Utilization of food, improving, 120, 127
Utopia, 6, 26, 28–29, 102; and technology, 21–22, 78–79, 85

Vasectomy, 35, 39–40
Vepex process, 122–23
Volcanoes, environmental effects of, 97

Weisberg, Barry, 27
Weiss, P., 34
Wertheim, W., 106
Western culture, predictions of, 103, 106
Westman, W., 24
White, Lynn, 25, 49–50
Woman's liberation, effects on pop-

ulation growth, 65, 126
Wordsworth, 70
Work and leisure, concepts of,
   45–46, 82, 100

Yablonsky, Lewis, 53, 100

Z.P.G. (zero population growth),
   9, 62